Studies in Systems, Decision and Control

Volume 349

The series "Studies in Systems, Decision and Control" (SSDC) covers both new developments and advances, as well as the state of the art, in the various areas of broadly perceived systems, decision making and control–quickly, up to date and with a high quality. The intent is to cover the theory, applications, and perspectives on the state of the art and future developments relevant to systems, decision making, control, complex processes and related areas, as embedded in the fields of engineering, computer science, physics, economics, social and life sciences, as well as the paradigms and methodologies behind them. The series contains monographs, textbooks, lecture notes and edited volumes in systems, decision making and control spanning the areas of Cyber-Physical Systems, Autonomous Systems, Sensor Networks, Control Systems, Energy Systems, Automotive Systems, Biological Systems, Vehicular Networking and Connected Vehicles, Aerospace Systems, Automation, Manufacturing, Smart Grids, Nonlinear Systems, Power Systems, Robotics, Social Systems, Economic Systems and other. Of particular value to both the contributors and the readership are the short publication timeframe and the world-wide distribution and exposure which enable both a wide and rapid dissemination of research output.

Indexed by SCOPUS, DBLP, WTI Frankfurt eG, zbMATH, SCImago.

All books published in the series are submitted for consideration in Web of Science.

More information about this series at http://www.springer.com/series/13304

Jinpeng Yu · Peng Shi · Jiapeng Liu

Intelligent Backstepping Control for the Alternating-Current Drive Systems

Springer

Jinpeng Yu
School of Automation
Qingdao University
Qingdao, China

Jiapeng Liu
School of Automation
Qingdao University
Qingdao, China

Peng Shi
School of Electrical
and Electronic Engineering
University of Adelaide
Adelaide, SA, Australia

ISSN 2198-4182 ISSN 2198-4190 (electronic)
Studies in Systems, Decision and Control
ISBN 978-3-030-67722-0 ISBN 978-3-030-67723-7 (eBook)
https://doi.org/10.1007/978-3-030-67723-7

This Springer imprint is published by the registered company Springer Nature Switzerland AG
The registered company address is: Gewerbestrasse 11, 6330 Cham, Switzerland

To Yumei and Jingchen

—Jinpeng Yu

To Fengmei, Lisa and Michael

—Peng Shi

To my parents and Huiying

—Jiapeng Liu

Preface

Modern electrical drives based on alternating current (AC) motors are intensively used in industrial applications and agriculture productions, such as steel mills, power plants. However, it is still challenging for us to achieve the perfect dynamic performance by controlling the AC motor because of its multi-variable, coupled and highly nonlinear dynamic model. The field-oriented control (FOC) and direct-torque control (DTC) are two of the most important developments in this field. Unfortunately, these control approaches suffer from sensitivity to the motor parameter variations and load disturbances. Stochastic disturbance has always been considered as a common source of instability of the AC motor control system. When the AC motor is working in a light load condition or running at a high speed, too many iron losses will be generated, which may create a negative impact on the control performance. In addition, the problem of "explosion of complexity" in the traditional backstepping control method will be inevitably arisen because of the continuous derivation of virtual control laws. Thus, the research on the intelligent control for the AC motor with uncertainty is attractive because of both theoretical and practical values.

This book focuses on the intelligent control design for both the induction motor (IM) and the permanent magnet synchronous motor (PMSM). The first chapter of this book introduces the research background of the AC motor, as well as dynamic models of both IM and PMSM. The general layout of the presentation of this book is divided into three parts. Part I proposes the intelligent controllers for the IM via backstepping approach. Part II focuses on the intelligent control design problems for the PMSM. These methodologies provide a framework for intelligent controller design, Lyapunov stability proof and performance analysis for AC motors. The main contents of Part I include the following: Chap. 2 is concerned with the problem of position tracking control for field-oriented IM with parameter uncertainties and load torque disturbance; Chap. 3 studies neural networks approximation-based command filtered adaptive control for the IM with input saturation; Chap. 4 addresses the discrete-time command filtered adaptive position tracking control problem for the IM via backstepping; Chap. 5 investigates the stochastic disturbances and input saturation problems for the IM drive systems; and

Chap. 6 is concerned with the adaptive fuzzy dynamic surface control for IM with iron losses. The main contents of Part II include the following: Chap. 7 is concerned with the speed tracking control problem of PMSM with parameter uncertainties and load torque disturbance; Chap. 8 studies the adaptive fuzzy backstepping position tracking control for the PMSM; Chap. 9 addresses the problem of neural-networks-based adaptive dynamic surface control for the PMSM with parameter uncertainties and load torque disturbance; Chap. 10 investigates the problem of discrete-time adaptive position tracking control for the interior PMSM based on fuzzy-approximation; Chap. 11 investigates adaptive fuzzy tracking control for the chaotic PMSM drive system via backstepping; Chap. 12 focuses on the problem of position tracking control for the chaotic PMSM drive system with parameter uncertainties. Finally, Chap. 13 in Part III summarizes the results of the book and discusses some future works.

This book is a research monograph, which provides valuable reference material for researchers who wish to explore the area of AC motor. In addition, the main contents of the book are also suitable for a one-semester graduate course.

Qingdao, China — Jinpeng Yu
Adelaide, Australia — Peng Shi
Qingdao, China — Jiapeng Liu
December 2020

Acknowledgements

The authors would like to thank numerous individuals who propose constructive comments, useful suggestions and wealth of ideas. Without them, this monograph could not have been completed. Special thanks go to Prof. Wenjie Dong from the University of Texas-Pan American, Prof. Bing Chen from Qingdao University, Prof. Haisheng Yu from Qingdao University and Prof. Chong Lin from Qingdao University, for their valuable suggestions, constructive comments and support.

Next, our acknowledgements go to many colleagues, who have offered support and encouragement throughout this research effort. In particular, we would like to acknowledge the contributions of Xuewei Mao from Qingdao University and Lin Zhao from Qingdao University. We also thank our students and their commentary. Finally, we would like to thank the editors at Springer for their professional and efficient handling of this project.

This work was partially supported by the National Natural Science Foundation of China (61573204, 61573203, 61973179), the Taishan Scholar Special Project Found (TSQN20161026), the Shandong Province Outstanding Youth Fund (ZR2011FQ012, ZR2015JL022), the National Key Research and Development Plan of China (2017YFB130503), the Science and Technology Project of College and University in Shandong Province (J11LG04), and the China Postdoctoral Science Foundation (2014T70620).

Contents

Notations and Acronyms

■	End of proof
♦	End of remark
$\in$	Belongs to
$\forall$	For all
$\sum$	Sum
$\mathbb{R}$	Field of real numbers
$\mathbb{R}^n$	Space of n-dimensional real vectors
lim	Limit
max	Maximum
min	Minimun
sup	Supremum
$\lambda_i(\cdot)$	ith eigen value of a matrix
$\lambda_{\min}(\cdot)$	Minimum eigen value of a matrix
$\lambda_{\max}(\cdot)$	Maximum eigen value of a matrix
$\lvert\cdot\rvert$	Euclidean vector norm
$\lvert\lvert\cdot\rvert\rvert$	Euclidean matrix norm (spectral norm)
FLS	Fuzzy logic system
FLC	Fuzzy logic control
NNs	Neural networks
PMSM	Permanent magnet synchronous maotor
IPMSM	Interior permanent magnet synchronous motor
IM	Induction motor
SMC	Sliding mode control
CFC	Command filter control
DSC	Dynamic surface control
RBF	Radial basis function

Matrices, if their dimensions are not explicitly stated, are assumed to be compatible for algebraic operations.

List of Figures

Chapter 1
Introduction

Modern electric drive of AC motors are the most spread used automatic electromechanical systems. AC motors including the IM and the PMSM are multivariable, high-order and strong coupling nonlinear systems [1, 2]. It is a research topic with theoretical significance and practical application value to study advanced control strategy and improve the dynamic and static performance of the AC drive system.

The AC motor is a high-order, strong coupling, multivariable, parameter time-varying nonlinear system [3]. The traditional control methods have not completely solved the performance problems such as time-varying motor parameters, uncertain load, unsuitable measurement of magnetic flux, low speed and zero speed.

A large number of research works have been done to improve the dynamic and static performance of the AC motor. Vaez-Zadeh and Jalali combined the field-oriented control and direct torque control [4] for high-performance AC motor in [5]. And a variety of AC motor controllers based on advanced nonlinear design techniques are successfully used to control the IM drives and references therein. Lin and Lee [6] introduced an adaptive backstepping control for a linear AC motor drive to track periodic reference inputs. Although this control strategy had good tracking performance and was insensitive to uncertainties, prior system knowledge was required in the control design. Wai et al. [7] developed a sliding-mode controller for field-oriented AC motor servo drive in which can overcome the common drawback of field-oriented control. Theoretically, the sliding motion is smooth if the switching frequency of a system is infinite. However, in practice, the switching frequency of a system is finite, thus chattering comes out along the sliding surface [8, 9]. Adaptive input-output linearizing control was proposed by Marino et al. [10]. Chiasson developed a new approach to dynamic feedback linearization control for an AC motor in [11]. But the employed method of feedback linearization requires the exact mathematical model, so the controller requires the desired dynamics to replace the system at the $d - q$ axis stator currents in [12]. Adaptive feedback linearization control was designed based on the air-gap flux model in [13] by Jeon, Baang and Choi with

J. Yu et al., *Intelligent Backstepping Control for the Alternating-Current Drive Systems*, Studies in Systems, Decision and Control 349,
https://doi.org/10.1007/978-3-030-67723-7_1

the rotor and stator resistances being unknown. Rashed et al. developed a nonlinear adaptive state feedback speed control for a voltage-fed IM with varying parameters in [14]. Yazdanpanah et al. [15] proposed an adaptive input-output feedback linearization and sliding mode control method to control torque and stator flux controller for AC motor. Therefore, the research of advanced control strategy for the AC motor drive system is still a research topic now.

Next, the dynamic models of the IM and the PMSM will be given in the next subsection.

1.1 Dynamic Mathematical Model for IM

To facilitate the theoretical analysis for IM, the following assumptions are given as [16, 17]:

(1) Ignore space harmonic, and set three-phase winding symmetry, the difference is 120 electrical degrees in space. The generated magnetomotive force is sinusoidal along the air gap;
(2) Ignore the saturation of the magnetic circuit, the self-inductance and mutual inductance of each winding are constant;
(3) Ignore core loss;
(4) The influence of frequency change and temperature change on winding resistance is not considered.

The physical model of the three-phase IM is shown in Fig. 1.1. The stator three-phase winding axes A, B and C are fixed in space, and the rotor winding axes a, b and c rotate with the rotor. The axis A is taken as the reference coordinate axis, and the electrical angle θ between the rotor a axis and the stator A axis is the spatial angular displacement variable.

The dynamic model of IM consists of voltage equation, flux equation, torque equation and motion equation [18].

The voltage equation is written in matrix form as

$$\begin{bmatrix} u_A \\ u_B \\ u_C \\ u_a \\ u_b \\ u_c \end{bmatrix} = \begin{bmatrix} R_s & 0 & 0 & 0 & 0 & 0 \\ 0 & R_s & 0 & 0 & 0 & 0 \\ 0 & 0 & R_s & 0 & 0 & 0 \\ 0 & 0 & 0 & R_r & 0 & 0 \\ 0 & 0 & 0 & 0 & R_r & 0 \\ 0 & 0 & 0 & 0 & 0 & R_r \end{bmatrix} \begin{bmatrix} i_A \\ i_B \\ i_C \\ i_a \\ i_b \\ i_c \end{bmatrix} + \frac{d}{dt} \begin{bmatrix} \psi_A \\ \psi_B \\ \psi_C \\ \psi_a \\ \psi_b \\ \psi_c \end{bmatrix}, \tag{1.1}$$

or written as

$$u = R_i + \frac{d\psi}{dt}, \tag{1.2}$$

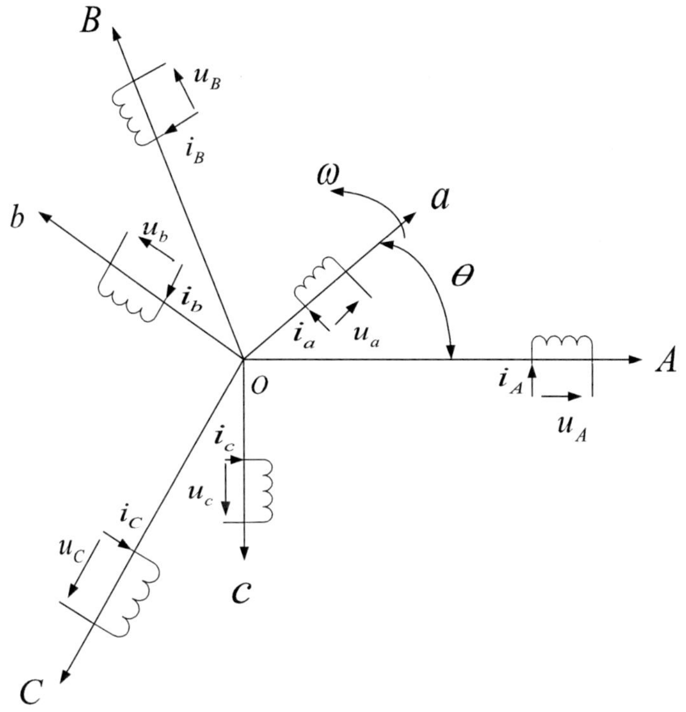

Fig. 1.1 Physical model of three-phase IM

where $u_A, u_B, u_C, u_a, u_b, u_c$ represent the instantaneous values of stator and rotor phase voltage, $i_A, i_B, i_C, i_a, i_b, i_c$ represent the instantaneous values of stator and rotor phase current, $\psi_A, \psi_B, \psi_C, \psi_a, \psi_b, \psi_c$ represent the full flux of each phase winding, and R_s, R_r represent the stator and rotor winding resistance.

The flux of each winding is the sum of its own self-induction flux and the mutual inductance flux of other windings. Therefore, the flux of six windings can be expressed as

$$\begin{bmatrix} \psi_A \\ \psi_B \\ \psi_C \\ \psi_a \\ \psi_b \\ \psi_c \end{bmatrix} = \begin{bmatrix} L_{AA} & L_{AB} & L_{AC} & L_{Aa} & L_{Ab} & L_{Ac} \\ L_{BA} & L_{BB} & L_{BC} & L_{Ba} & L_{Bb} & L_{Bc} \\ L_{CA} & L_{CB} & L_{CC} & L_{Ca} & L_{Cb} & L_{Cc} \\ L_{aA} & L_{aA} & L_{aC} & L_{aa} & L_{ab} & L_{ac} \\ L_{bA} & L_{bA} & L_{bC} & L_{ba} & L_{bb} & L_{bc} \\ L_{cA} & L_{cA} & L_{cC} & L_{ca} & L_{cb} & L_{cc} \end{bmatrix} \begin{bmatrix} i_A \\ i_B \\ i_C \\ i_a \\ i_b \\ i_c \end{bmatrix}, \tag{1.3}$$

or written as

$$\psi = Li, \tag{1.4}$$

where L is the 6×6 inductance matrix, the diagonal element L_{AA}, L_{BB}, L_{CC}, L_{aa}, L_{bb}, L_{cc} are the self-inductance of each winding, and the rest is the mutual inductance between corresponding windings.

If the flux equation is substituted into the voltage equation, the expanded voltage equation can be obtained:

$$\begin{aligned} u &= Ri + \tfrac{d}{dt}(Li) = Ri + L\tfrac{di}{dt} + \tfrac{dL}{dt}i \\ &= Ri + L\tfrac{di}{dt} + \tfrac{dL}{d\theta}\omega i, \end{aligned} \tag{1.5}$$

where $L\frac{di}{dt}$ is the electromotive force of the transformer caused by the current change, $\frac{dL}{d\theta}\omega i$ is the rotating electromotive force which is proportional to the rotating speed caused by the relative position change of the stator and rotor.

Furthermore, the torque equation can be written as

$$\begin{aligned} T_e = \; & -n_p L_{ms}[(i_A i_a + i_B i_b + i_C i_c)\sin\theta + (i_A i_b + i_B i_c + i_C i_a)\sin(\theta + 120^\circ) \\ & + (i_A i_c + i_B i_a + i_C i_b)\sin(\theta - 120^\circ)]. \end{aligned} \tag{1.6}$$

The corresponding equation of motion is

$$\frac{J}{n_p}\frac{d\omega}{dt} = T_e - T_L, \tag{1.7}$$

where J is the moment of inertia of the unit, and T_L is the load torque including friction resistance torque and elastic torque.

The mathematical expression of the asynchronous motor dynamic model angle equation of asynchronous motor can be expressed as

$$\frac{d\theta}{dt} = \omega, \tag{1.8}$$

Combine the equation of motion (1.7), it has

$$\frac{d\omega}{dt} = \frac{n_p}{J}(T_e - T_L), \tag{1.9}$$

and the expanded voltage equation

$$L\frac{di}{dt} = -Ri - \frac{dL}{d\theta}\omega i + u. \tag{1.10}$$

The eighth order differential equations with state variables $[\theta \; \omega \; i_A \; i_B \; i_C \; i_a \; i_b \; i_c]^T$ and input variables $[u_A \; u_B \; u_C \; T_L]^T$ are obtained, in which T_L is the disturbance input.

The structure diagram of the vector control system of three-phase current closed-loop control is shown in Fig. 1.2. The given values of the two components of stator

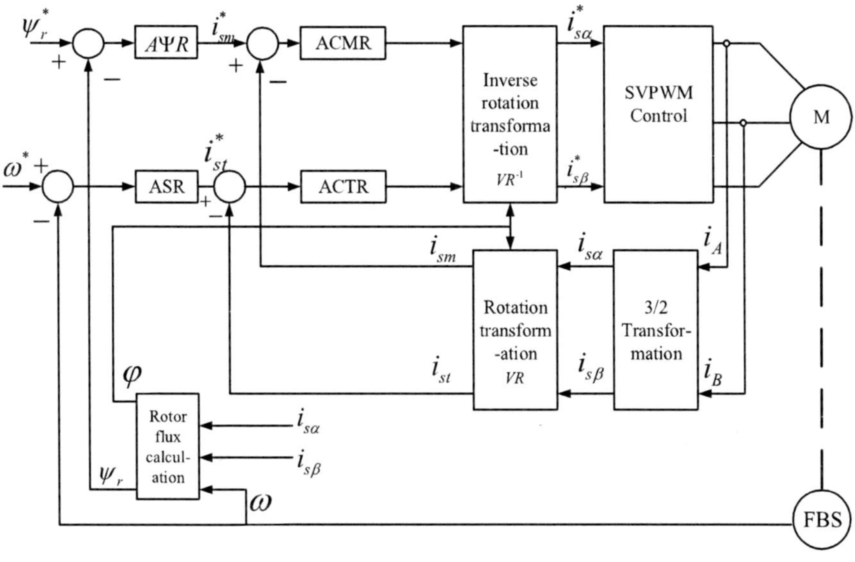

Fig. 1.2 Traditional structure diagram of the vector control system of IM

current i^*_{sm} and i^*_{st} are converted by 2/3 to get the sum of the given values of three-phase current i^*_A, i^*_B and i^*_C. The current controlled PWM inverter is used to complete the current closed-loop control in the three-phase stator coordinate system.

Under the assumptions of equal mutual inductance and a linear magnetic circuit and through the field-oriented transformation, a fifth-order IM, which includes both the electrical and mechanical dynamics, can be described in the well known $(d-q)$ frame as follows:

$$\begin{cases} \dfrac{d\theta}{dt} = \omega, \\ \dfrac{d\omega}{dt} = \dfrac{n_p L_m}{L_r J}\psi_d i_q - \dfrac{T_L}{J}, \\ \dfrac{di_q}{dt} = -\dfrac{L_m^2 R_r + L_r^2 R_s}{\sigma L_s L_r^2} i_q - \dfrac{n_p L_m}{\sigma L_s L_r}\omega\psi_d - n_p\omega i_d - \dfrac{L_m R_r}{L_r}\dfrac{i_q i_d}{\psi_d} + \dfrac{1}{\sigma L_s}u_q, \\ \dfrac{d\psi_d}{dt} = -\dfrac{R_r}{L_r}\psi_d + \dfrac{L_m R_r}{L_r} i_d, \\ \dfrac{di_d}{dt} = -\dfrac{L_m^2 R_r + L_r^2 R_s}{\sigma L_s L_r^2} i_d + \dfrac{L_m R_r}{\sigma L_s L_r^2}\psi_d + n_p\omega i_q + \dfrac{L_m R_r}{L_r}\dfrac{i_q^2}{\psi_d} + \dfrac{1}{\sigma L_s}u_d, \end{cases}$$

where $\sigma = 1 - \frac{L_m^2}{L_s L_r}$. θ, ω, L_m, n_p, J, T_L and ψ_d denote the rotor position, rotor angular velocity, mutual inductance, pole pairs, inertia, load torque and rotor flux linkage. i_d and i_q stand for the $d-q$ axis currents. u_d and u_q are the $d-q$ axis voltages. R_s and L_s mean the resistance, inductance of the stator. R_r and L_r denote the resistance, inductance of the rotor. For simplicity, the following notations are introduced:

$$
\begin{aligned}
&x_1 = \theta, x_2 = \omega, x_3 = i_q, x_4 = \psi_d, x_5 = i_d,\\
&a_1 = \frac{n_p L_m}{L_r}, b_1 = -\frac{L_m^2 R_r + L_r^2 R_s}{\sigma L_s L_r^2},\\
&b_2 = -\frac{n_p L_m}{\sigma L_s L_r}, b_3 = n_p, b_4 = \frac{L_m R_r}{L_r}, b_5 = \frac{1}{\sigma L_s},\\
&c_1 = -\frac{R_r}{L_r}, d_2 = \frac{L_m R_r}{\sigma L_s L_r^2}.
\end{aligned}
$$

By using these notations, the dynamic model of IM drivers can be described by the following differential equations:

$$
\begin{cases}
\dot{x}_1 = x_2,\\
\dot{x}_2 = \frac{a_1}{J} x_3 x_4 - \frac{T_L}{J},\\
\dot{x}_3 = b_1 x_3 + b_2 x_2 x_4 - b_3 x_2 x_5 - b_4 \frac{x_3 x_5}{x_4} + b_5 u_q,\\
\dot{x}_4 = c_1 x_4 + b_4 x_5,\\
\dot{x}_5 = b_1 x_5 + d_2 x_4 + b_3 x_2 x_3 + b_4 \frac{x_3^2}{x_4} + b_5 u_d,
\end{cases}
\tag{1.11}
$$

where u_q and u_d are the scalar control signals.

1.2 Dynamic Mathematical Model for PMSM

To facilitate the theoretical analysis for PMSM, make the following assumptions[19]:

(1) Ignore space harmonic, set three-phase winding symmetry, space difference is 120°. The generated magnetomotive force is sinusoidal along the air gap;
(2) Ignore the saturation of the magnetic circuit, the self-inductance and mutual inductance of each winding are constant;
(3) Ignore core loss;
(4) The influence of frequency change and temperature change on winding resistance is not considered;

The physical model of the synchronous motor with damping winding is shown in Fig. 1.3. The axis A, B and C of the stator three-phase winding are static, u_A, u_B, u_C are the three-phase stator voltage, i_A, i_B, i_C are the three-phase stator current, the rotor rotates at angular speed. The excitation winding on the rotor flows through the excitation current I_f under the supply of excitation voltage U_f. The axis along the excitation pole is d axis, and q axis is orthogonal to d axis. The coordinate system $d-q$ is fixed on the rotor and rotates synchronously with the rotor. The angle between d axis and A axis is variable θ_r. i_{rd} and i_{rq} are the d axis and q axis currents of the damping winding.

The voltage matrix equation of PMSM is

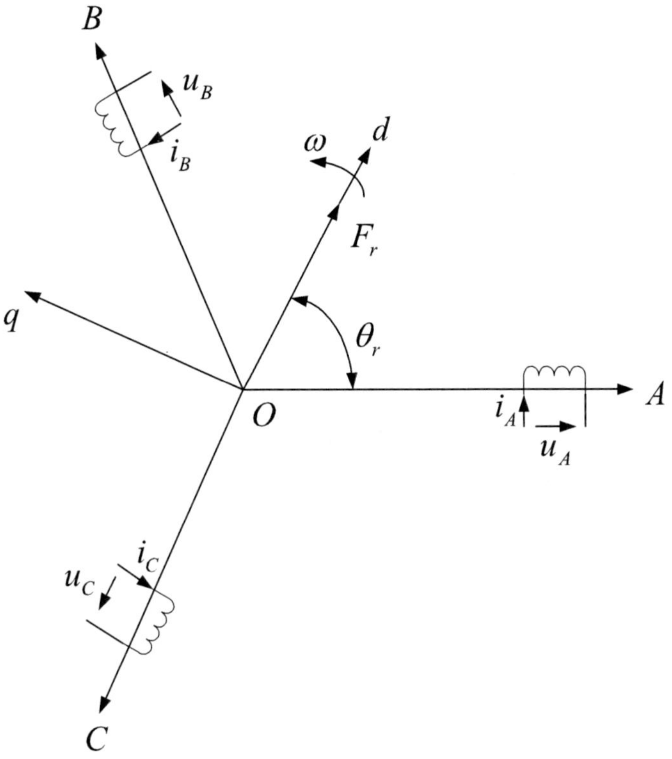

Fig. 1.3 Physical model of PMSM

$$\begin{bmatrix} u_{sd} \\ u_{sq} \\ U_f \\ 0 \\ 0 \end{bmatrix} = \begin{bmatrix} R_s & -\omega L_{sq} & 0 & 0 & -\omega L_{mq} \\ \omega L_{sd} & R_s & \omega L_{md} & \omega L_{md} & 0 \\ 0 & 0 & R_f & 0 & 0 \\ 0 & 0 & 0 & R_{rd} & 0 \\ 0 & 0 & 0 & 0 & R_{rq} \end{bmatrix} \begin{bmatrix} i_{sd} \\ i_{sq} \\ I_f \\ i_{rd} \\ i_{rq} \end{bmatrix} + \begin{bmatrix} L_{sd} & 0 & L_{md} & L_{md} & 0 \\ 0 & L_{sq} & 0 & 0 & L_{mq} \\ L_{md} & 0 & L_f & L_{md} & 0 \\ L_{md} & 0 & L_{md} & L_{rd} & 0 \\ 0 & L_{mq} & 0 & 0 & L_{rq} \end{bmatrix} \frac{d}{dt} \begin{bmatrix} i_{sd} \\ i_{sq} \\ I_f \\ i_{rd} \\ i_{rq} \end{bmatrix}, \tag{1.12}$$

where L_{sd} is the equivalent two-phase stator winding d-axis self-inductance, L_{sq} is the equivalent two-phase stator winding q-axis self-inductance, L_{md} is mutual inductance between d-axis stator and rotor winding, L_{mq} is mutual inductance between q-axis stator and rotor winding, L_f is self-inductance of excitation winding, L_{rd} is self-inductance of d-axis damping winding, L_{rq} is self-inductance of q-axis damping winding group.

The corresponding equation of motion is

$$\frac{d\omega}{dt} = \frac{n_p}{J}(T_e - T_L) = \frac{n_p^2}{J}\left[L_{md}I_f i_{sq} + \left(L_{sd} - L_{sq}\right) i_{sd} i_{sq} \right.$$
$$\left. + \left(L_{md} i_{rd} i_{sq} - L_{mq} i_{rq} i_{sd}\right)\right] - \frac{n_p}{J} T_L. \quad (1.13)$$

The mathematical dynamic model is

$$\begin{bmatrix} u_{sd} \\ u_{sq} \\ U_f \end{bmatrix} = \begin{bmatrix} R_s & -\omega L_s & 0 \\ \omega L_s & R_s & \omega L_m \\ 0 & 0 & R_f \end{bmatrix} \begin{bmatrix} i_{sd} \\ i_{sq} \\ I_f \end{bmatrix} + \begin{bmatrix} L_s & 0 & L_m \\ 0 & L_s & 0 \\ L_m & 0 & L_f \end{bmatrix} \frac{\mathrm{d}}{\mathrm{d}t} \begin{bmatrix} i_{sd} \\ i_{sq} \\ I_f \end{bmatrix}, \quad (1.14)$$

$$\frac{\mathrm{d}\omega}{\mathrm{d}t} = \frac{n_p}{J}(T_e - T_L) = \frac{n_p^2}{J} L_m I_f i_{sq} - \frac{n_p}{J} T_L. \quad (1.15)$$

The model of PMSM can be described within $d-q$ frame through the Park transformation as follows [20]:

$$\begin{aligned} \frac{d\theta}{dt} &= \omega, \\ J\frac{d\omega}{dt} &= \frac{3}{2} n_p \left[(L_d - L_q) i_d i_q + \Phi i_q\right] - B\omega - T_L, \\ L_d \frac{di_d}{dt} &= -R_s i_d + n_p \omega L_q i_q + u_d, \\ L_q \frac{di_q}{dt} &= -R_s i_q - n_p \omega L_d i_d - n_p \omega \Phi + u_q, \end{aligned}$$

where θ is the rotor position, ω denotes the rotor angular velocity, i_d and i_q stand for the $d-q$ axis currents, u_d and u_q are the $d-q$ axis voltages, R_s is the stator resistance, L_d and L_q are the $d-q$ axis stator inductors, n_p is the pole pair, J means the rotor moment of inertia, B is the viscous friction coefficient, T is the electromagnetism torque, T_L is the load torque and Φ denotes magnet flux linkage. To simplify the above model, the following notations are introduced:

$$\begin{aligned} & x_1 = \theta, x_2 = \omega, x_3 = i_q, x_4 = i_d, a_1 = \frac{3n_p\Phi}{2}, \\ & a_2 = \frac{3n_p(L_d - L_q)}{2}, b_1 = -\frac{R_s}{L_q}, b_2 = -\frac{n_p L_d}{L_q}, \\ & b_3 = -\frac{n_p \Phi}{L_q}, b_4 = \frac{1}{L_q}, c_1 = -\frac{R_s}{L_d}, c_2 = \frac{n_p L_q}{L_d}, c_3 = \frac{1}{L_d}. \end{aligned}$$

By using these notations, the dynamic model of the PMSM can be described by the following differential equations:

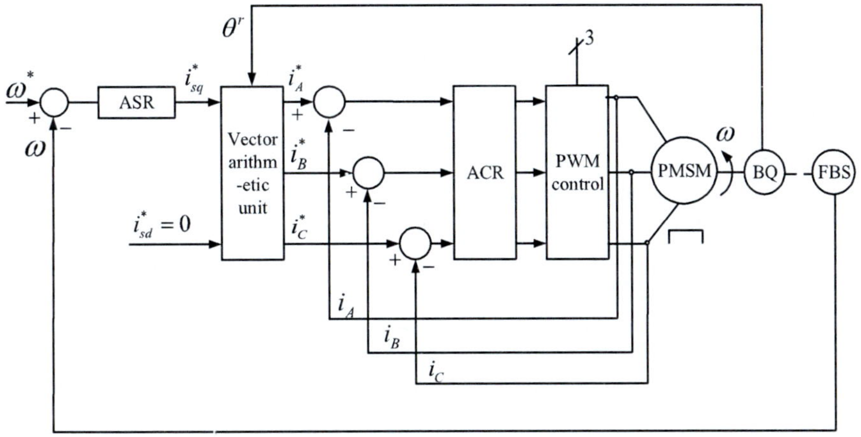

Fig. 1.4 Traditional vector control system of $i_{sd} = 0$ PMSM oriented by rotor flux

$$\begin{cases} \dot{x}_1 = x_2, \\ \dot{x}_2 = \frac{a_1}{J}x_3 + \frac{a_2}{J}x_3x_4 - \frac{B}{J}x_2 - \frac{T_L}{J}, \\ \dot{x}_3 = b_1x_3 + b_2x_2x_4 + b_3x_2 + b_4u_q, \\ \dot{x}_4 = c_1x_4 + c_2x_2x_3 + c_3u_d. \end{cases} \tag{1.16}$$

The principle block diagram of the vector control system of PMSM based on rotor flux linkage orientation and $i_{sd} = 0$ is shown in Fig. 1.4, like a DC motor speed control system. The output of speed regulator ASR is proportional to the stator current given value of electromagnetic torque.

1.3 Outline of the Book

The general layout of the presentation of this book is divided into three parts. Part I: intelligent backstepping control schemes of IM, Part II: intelligent backstepping control schemes of PMSM, and Part III: summary of the book.

This chapter introduced the research background and significance of the AC motors, as well as the challenge of designing high-quality AC system control schemes. Secondly, an overview of the work in this book has been provided. Additionally, the physical models of IM and PMSM are introduced, and detailed mathematical modeling processes are described.

Part I focuses on the stability analysis and control design for IM. Part I which begins with Chap. 2 consists of five chapters as follows.

Chapter 2 focuses on the position tracking control for the IM with parameter uncertainties and load torque disturbance. The fuzzy logic system (FLS) is used to approximate the nonlinearities and an adaptive backstepping technique is employed to construct controllers. The proposed adaptive fuzzy controllers guarantee the tracking error converges to a small neighborhood of the origin. Simulation results are analyzed to prove the effectiveness of the proposed scheme.

Chapter 3 considers the problem of input saturation in the IM. The neural networks are utilized to approximate the nonlinearities, and the command filtering technology is used to deal with the "explosion of complexity" problem caused by the derivative of virtual controllers in the conventional backstepping design. The compensating signals are further exploited to get rid of the drawback caused by the dynamics of filter technology. It is verified that the adaptive neural controller guarantees that the tracking error can converge to a small neighborhood of the origin. At last, our results are given to show the effectiveness and advantages of the proposed techniques.

Chapter 4 investigates the IM discrete-time drive systems with parameter uncertainties and load disturbance. First, the Euler method is used to describe the discrete-time dynamic mathematical model of IM. Next, the neural networks technique is employed to approximate the unknown nonlinear functions. Furthermore, the "explosion of complexity" problem and noncausal problem, which emerged in traditional backstepping design, are eliminated by command filtered control technique. Simulation results are analyzed to prove the effectiveness of the proposed approach.

Chapter 5 considers the stochastic disturbances and input saturation problems in the IM drive system. Firstly, the FLS is employed to cope with the stochastic nonlinear functions in the IM drive system. Secondly, the quartic Lyapunov function is selected as the stochastic Lyapunov function and an adaptive backstepping method is used to design controllers. Then the command filtered technology is utilized to deal with the "explosion of complexity" in conventional backstepping, and the filtering error is eliminated by designing compensating signal. Finally, the effectiveness and superiority of the proposed method are demonstrated by simulation results.

Chapter 6 investigates the dynamic surface control (DSC) method combined with adaptive fuzzy backstepping technology for IM with iron losses in electric vehicle drive systems. The DSC is utilized to overcome the "explosion of complexity" problem of classical backstepping. The FLS is used to approximate unknown nonlinear functions and an adaptive backstepping scheme is employed to design controllers. The proposed control method can guarantee all the closed-loop signals are bounded. Simulation results illustrate its effectiveness.

Part II focuses on the stability analysis and control design for PMSM. Part II which begins with Chap. 7 consists of six chapters as follows.

Chapter 7 studies the speed tracking control problem of PMSM with parameter uncertainties and load torque disturbance. The FLS is used to approximate nonlinearities and an adaptive backstepping technique is employed to construct controllers. The proposed controller guarantees the tracking error converges to a small neighborhood of the origin and achieves good tracking performance. Simulation results clearly show that the proposed control scheme can track the position reference signal generated by a reference model successfully under parameter uncertainties and load torque disturbance without singularity and overparameterization.

Chapter 8 is concerned with the position tracking control problem of PMSM with parameter uncertainties and load torque disturbance. Fuzzy logic systems are used to approximate nonlinearities and an adaptive backstepping technique is employed to construct controllers. The proposed adaptive fuzzy controllers guarantee that the tracking error converges to a small neighborhood of the origin. Compared with the conventional backstepping method, the proposed fuzzy controllers' structure is very simple and easy to be implemented in practice. The simulation results illustrate the effectiveness of the proposed results.

Chapter 9 considers the problem of neural-networks-based adaptive DSC for the PMSM with parameter uncertainties and load torque disturbance. First, neural networks are used to approximate the unknown and nonlinear functions of the PMSM drive system and a novel adaptive DSC is constructed to avoid the "explosion of complexity" in the backstepping design. The number of adaptive parameters required is reduced to only one, and the designed neural controllers' structure is much simpler than some existing results in the literature. Then, simulations are given to illustrate the effectiveness and potential of the new design technique.

Chapter 10 is concerned with the problem of discrete-time adaptive position tracking control for an interior permanent magnet synchronous motor (IPMSM). The FLS is used to approximate the nonlinearities of the discrete-time IPMSM drive system which is derived by direct discretization using the Euler method, and a discrete-time fuzzy position tracking controller is designed via backstepping. Simulation results illustrate the effectiveness and the potentials of the theoretic results obtained.

Chapter 11 investigates an adaptive fuzzy control method to suppress chaos in the PMSM drive system. An adaptive fuzzy backstepping technique is employed to construct controllers. The simulation results are given to illustrate the effectiveness of the proposed control scheme.

Chapter 12 studies the problem of position tracking control for the chaotic PMSM drive system with parameter uncertainties. The FLS is used to approximate the nonlinearities and an adaptive backstepping technique is employed to construct controllers. The proposed adaptive fuzzy controllers guarantee that the tracking error converges to a small neighborhood of the origin. Simulation results are given to show the effectiveness of the proposed control scheme.

Part III summarizes the results of the book.

Chapter 13 summarizes the results of the book and then proposes some related topics for future research work.

References

1. Lee, J.J.: Design of multivariable variable structure system for nonlinear time-varying systems using nonlinear switching surfaces. Electron. Lett. **27**(23), 2111–2113 (2002)
2. Minoru, K., Junya, K., Nobuo, T.: Performance comparison between a permanent magnet synchronous motor and an induction motor as a traction motor for high speed train. IEEE Trans. Ind. Appl. **126**(2), 168–173 (2006)
3. Yu, J.P., Yu, H.S., Gao, J.W., Cheng, X.Q., Qin, Y.: Chaos control of permanent magnet synchronous motors based on fuzzy-approximation. Complex Syst. Complex. Sci. **10**(4), 86–91 (2013)
4. Prasad, D., Panigrahi, B.P., SenGupta, S.: Digital simulation and hardware implementation of a simple scheme for direct torque control of induction motor. Energy Convers. Manag. **49**, 687–697 (2008)
5. Vaez-Zadeh, S., Jalali, E.: Combined vector control and direct torque control method for high performance induction motor drives. Energy Convers. Manag. **48**, 3095–31001 (2007)
6. Lin, F.J., Lee, C.C.: Adaptive backstepping control for linear induction motor drive to track periodic references. Proc. Inst. Electr. Eng. Electric Power Appl. **147**(6), 449–458 (2000)
7. Wai, R.J., Lin, K.M., Lin, C.Y.: Total sliding-mode speed control of fieldoriented induction motor servo drive. In: Proceedings of the 5th Asian Control Conference, Australia (2004)
8. Liu, J.K., Sun, F.C.: Research and development on theory and algorithms of sliding mode control. IET Control Theory Appl. **24**(3), 407–418 (2007)
9. Sarwer, M.G., Rafiqn, Md.A., Data, M., Ghosh, B.C., Komada, S.: Chattering free neuro-sliding mode control of DC drives. In: IEEE Proceedings of International Conference on Power Electronics and Drivers Systems, pp. 1101–1106 (2005)
10. Marino, R., Peresada, S., Valigi, P.: Adaptive input-output linearizing control of induction motors. IEEE Trans. Autom. Control **38**(2), 208–221 (1993)
11. Chiasson, J.: A new approach to dynamic feedback linearization control of an induction motor. IEEE Trans. Autom. Control **43**(2), 391–397 (1998)
12. Isidori, A.: Nonlinear Control Systems, 3rd edn. Springer, Berlin (1995)
13. Jeon, S.H., Baang, D., Choi, J.Y.: Adaptive feedback linearization control based on airgap flux model for induction motors. In: 30th Annual Conference of the IEEE Industrial Electronics Society, Korea, pp. 1099–1104 (2004)
14. Rashed, M., Maconnell, P.F.A., Stronach, A.F.: Nonlinear adaptive state feedback speed control of a voltage-fed induction motor with varying parameters. IEEE Trans. Ind. Appl. **42**(3), 723–732 (2006)
15. Yazdanpanah, R., Soltani, J., Markadeh, G.R.: Arab: nonlinear torque and stator flux controller for induction motor drive based on adaptive input-output feedback linearization and sliding mode control. Energy Convers. Manag. **49**, 541–550 (2008)
16. Marino, R., Tomei, P., Verrelli, C.M.: An adaptive tracking control from current measurements for induction motors with uncertain load torque and rotor resistance. Automatica **44**(10), 2593–2599 (2008)
17. Pei, W.H., Zhang, C.H., Li, K., Cui, N.X.: Hamilton system modeling and passive control for induction motor of electric vehicles by considering iron losses. IET Control Theory Appl. **28**(6), 869–873 (2011)
18. Fu, C., Zhao, L., Yu, J.P., Yu, H.S.: Neural network-based command filtered control for induction motors with input saturation. IET Control Theory Appl. **11**(15), 2636–2642 (2017)
19. Qi, L., Shi, H.B.: Adaptive position tracking control of permanent magnet synchronous motor based on RBF fast terminal sliding mode control. Neurocomputing **115**(4), 23–30 (2013)
20. Sangsefidi, Y., Ziaeinejad, S., Mehrizi-Sani, A., Pairodin-Nabi, H., Shoulaie, A.: Estimation of stator resistance in direct torque control synchronous motor drives. IEEE Trans. Energy Convers. **30**(2), 624–634 (2015)

Part I
Induction Motor

Chapter 2
Position Tracking Control of IM via Adaptive Fuzzy Backstepping

This chapter considered the problem of position tracking control for field-oriented induction motor with parameter uncertainties and load torque disturbance. Fuzzy logic systems are employed to approximate the nonlinear functions and an adaptive backstepping method is used to constitute controllers. The proposed adaptive fuzzy controllers ensure that the tracking error converges to a small neighborhood of the origin. Compared with the traditional backstepping, the devised fuzzy controllers' structure is very simple. The simulation results illustrate the effectiveness of the proposed control strategy.

2.1 Introduction

Modern electric drive based on induction motor (IM) has been widely used in industrial applications for its advantages of simple structure, low cost, high reliability and durability. In order to obtain better control performance, advanced motion control technology is deeply studied [1–4]. IM is very sensitive to the motor parameter variations and load disturbances [5]. The adaptive backstepping method is a newly developed technique to control the nonlinear systems with parameter uncertainty, and a lot of significant results have been obtained [6–9].

In recent years, fuzzy logic control (FLC) [10–12] has been found as one of the popular tools in functional approximations. Therefore, an FLC can be used to handle uncertain information, furthermore, be applied to control these systems which are ill-defined or too complicate to have a mathematical model. Classically, fuzzy variables have been adjusted by expert knowledge and trial and error. It provides an effective way to design a control system that is one of the important applications in control engineering [13, 14].

J. Yu et al., *Intelligent Backstepping Control for the Alternating-Current Drive Systems*,
Studies in Systems, Decision and Control 349,
https://doi.org/10.1007/978-3-030-67723-7_2

In this chapter, an adaptive fuzzy control approach with the field orientation transformation is proposed for position tracking control of IM drive system via backstepping. During the controller design process, FLS is used to approximate the nonlinearities, adaptive technique and backstepping are employed to construct fuzzy controllers. This means that the indeterministic parameters are taken into account, no regression matrices need to be found and the problem of "explosion of terms" is overcome. Thus, the major problems with traditional backstepping are cured. To verify the advantage of the proposed control method, a comparison between the two methodologies was studied. Moreover, the proposed controllers guarantee that the tracking error converges to a small neighborhood of the origin and all the closed-loop signals are bounded. The simulation results are provided to demonstrate the effectiveness and robustness against the parameter uncertainties and load disturbances.

2.2 Mathematical Model of the IM Drive System and Preliminaries

From (1.11), the dynamic model of IM drivers [15] can be described by the following differential equations:

$$
\begin{aligned}
\dot{x}_1 &= x_2, \\
\dot{x}_2 &= \frac{a_1}{J} x_3 x_4 - \frac{T_L}{J}, \\
\dot{x}_3 &= b_1 x_3 + b_2 x_2 x_4 - b_3 x_2 x_5 - b_4 \frac{x_3 x_5}{x_4} + b_5 u_q, \\
\dot{x}_4 &= c_1 x_4 + b_4 x_5, \\
\dot{x}_5 &= b_1 x_5 + d_2 x_4 + b_3 x_2 x_3 + b_4 \frac{x_3^2}{x_4} + b_5 u_d.
\end{aligned} \tag{2.1}
$$

The control objective is to design an adaptive fuzzy controller such that the state variable x_i $(i = 1, 4)$ follows the given reference signal x_{id} and all the closed-loop signals are bounded. For this purpose, we adopt the singleton fuzzifier, product inference, and the center-defuzzifier to deduce the following fuzzy rules:

$$R_i : \text{IF } x_1 \text{ is } F_1^i \text{ and...and } x_n \text{ is } F_n^i \text{ THEN } y \text{ is } B^i \; (i = 1, 2, ..., N),$$

where $x = [x_1, ..., x_n]^T \in R^n$, and $y \in R$ are the input and output of the fuzzy system, respectively, F_i^j and B^i are fuzzy sets in R. The fuzzy inference engine performs a mapping from fuzzy sets in R^n to fuzzy set in R based on the IF-THEN rules in the fuzzy rule base and the compositional rule of inference. The fuzzifier maps a crisp point $x = [x_1, ..., x_n]^T \in R^n$ into a fuzzy set A_x in R. The defuzzifier maps a fuzzy set in R to a crisp point in R. Since the strategy of singleton fuzzification, center-average defuzzification and product inference is used, the output of the fuzzy system can be formulated as

$$y(x) = \frac{\sum_{j=1}^{N} W_j \prod_{i=1}^{n} \mu_{F_i^j}(x_i)}{\sum_{j=1}^{N} [\prod_{i=1}^{n} \mu_{F_i^j}(x_i)]},$$

where W_j is the point at which fuzzy membership function $\mu_{B^j}(W_j)$ achieves its maximum value, and it is further assumed that $\mu_{B^j}(W_j) = 1$. Let $p_j(x) = \frac{\prod_{i=1}^{n} \mu_{F_i^j}(x_i)}{\sum_{j=1}^{N} [\prod_{i=1}^{n} \mu_{F_i^j}(x_i)]}$, $S(x) = [p_1(x), p_2(x), ..., p_N(x)]^T$ and $W = [W_1, ..., W_N]^T$, then the fuzzy logic system above can be rewritten as

$$y(x) = W^T S(x). \tag{2.2}$$

If all memberships are taken as Gaussian functions, then the following lemma holds.

Lemma 2.1 *[16] Let $f(x)$ be a continuous function defined on a compact set Ω. Then for any scalar $\varepsilon > 0$, there exists a fuzzy logic system in the form (2.2) such that*

$$\sup_{x \in \Omega} |f(x) - y(x)| \leq \varepsilon.$$

2.3 Adaptive Fuzzy Controller Design with Backstepping

In this section, we will devise a control method for the IM system. The system (2.1) leads a simplified system structure with two approximately decoupled subsystems, namely, the subsystem with state variables (x_1, x_2, x_3) and control signal u_q, and the subsystem with (x_4, x_5) as state variables and u_d as the control input. The backstepping design procedure consists of 5 steps. At each design step, a virtual control function α_i $(i = 1, 2, 3)$ will be constructed by using a suitable Lyapunov function. Finally, the real controller is constructed to control the system.

Step 1: For the first subsystem, define the tracking error variable as $z_1 = x_1 - x_{1d}$. From the first differential equation of (2.1), it has $\dot{z}_1 = x_2 - \dot{x}_{1d}$.

Choose Lyapunov function candidate as $V_1 = \frac{1}{2}z_1^2$, then the time derivative of V_1 can be obtained as

$$\dot{V}_1 = z_1\dot{z}_1 = z_1(x_2 - \dot{x}_{1d}). \tag{2.3}$$

Construct the virtual control law α_1 as

$$\alpha_1(x_1, x_{1d}, \dot{x}_{1d}) = -k_1 z_1 + \dot{x}_{1d}, \tag{2.4}$$

with $k_1 > 0$ being a design parameter. By employing (2.4), (2.3) can be rewritten of the following form:

$$\dot{V}_1 = -k_1 z_1^2 + z_1 z_2,$$

with $z_2 = x_2 - \alpha_1$.

Step 2: Differentiating z_2 gives

$$\dot{z}_2 = \dot{x}_2 - \dot{\alpha}_1 = \frac{a_1}{J} x_3 x_4 - \frac{T_L}{J} - \dot{\alpha}_1. \tag{2.5}$$

Now, select the Lyapunov function candidate as $V_2 = V_1 + \frac{J}{2} z_2^2$. Obviously, the time derivative of V_2 is given by

$$\dot{V}_2 = \dot{V}_1 + \frac{J}{2} z_2 \dot{z}_2 = -k_1 z_1^2 + z_2(z_1 + a_1 x_3 x_4 - T_L - J\dot{\alpha}_1). \tag{2.6}$$

Remark 2.2 In this chapter owing to the parameters T_L being bounded in practice system, we assume the T_L is unknown, but its upper bound is $d > 0$, which may be unknown, namely, $0 \leq T_L \leq d$. ♦

Obviously, $z_2 T_L \leq \frac{1}{2\varepsilon_2^2} z_2^2 + \frac{1}{2}\varepsilon_2^2 d^2$, where ε_2 is an arbitrary small positive constant. Then the time derivative of V_2 satisfies the following inequality.

$$\dot{V}_2 \leq -k_1 z_1^2 + z_2(z_1 + \frac{1}{2\varepsilon_2^2} z_2 + a_1 x_3 x_4 - J\dot{\alpha}_1) + \frac{1}{2}\varepsilon_2^2 d^2. \tag{2.7}$$

Since J is unknown, it cannot be used to construct the control signal. Thus, let $\hat{J}$ be the estimation of J. The corresponding adaptation laws will be specified later. The virtual control α_2 is constructed as

$$\begin{aligned} \alpha_2(Z_2) &= \frac{1}{a_1 x_4}(-\bar{k}_2 z_2 - \frac{1}{2\varepsilon_2^2} z_2 - z_1 + \hat{J}\dot{\alpha}_1) \\ &= \frac{1}{a_1 x_4}(-k_2 z_2 - z_1 + \hat{J}\dot{\alpha}_1), \end{aligned} \tag{2.8}$$

where $k_2 = \bar{k}_2 + \frac{1}{2\varepsilon_2^2} > 0$ is a positive design parameter and $Z_2 = [x_1, x_2, x_{1d}, \dot{x}_{1d}, \ddot{x}_{1d}, \hat{J}]^T$. Adding and subtracting α_2 in the bracket in (2.7) shows that

$$\dot{V}_2 = -k_1 z_1^2 - k_2 z_2^2 + a_1 z_2 z_3 x_4 + z_2(\hat{J} - J)\dot{\alpha}_1 + \frac{1}{2}\varepsilon_2^2 d^2, \tag{2.9}$$

with $z_3 = x_3 - \alpha_2$.

Step 3: Differentiating z_3 results in the following differential equation.

$$\dot{z}_3 = \dot{x}_3 - \dot{\alpha}_2 = b_1 x_3 + b_2 x_2 x_4 - b_3 x_2 x_5 - b_4 \frac{x_3 x_5}{x_4} + b_5 u_q - \dot{\alpha}_2.$$

Choose the Lyapunov function candidate as $V_3 = V_2 + \frac{1}{2}z_3^2$. Furthermore, differentiating V_3 yields

$$\begin{aligned}\dot{V}_3 &= \dot{V}_2 + z_3\dot{z}_3 = \dot{V}_2 + z_3(b_1x_3 + b_2x_2x_4 - b_3x_2x_5 - b_4\frac{x_3x_5}{x_4} + b_5u_q - \dot{\alpha}_2)\\ &= -k_1z_1^2 - k_2z_2^2 + \frac{1}{2}\varepsilon_2^2d^2 + z_2(\hat{J} - J)\dot{\alpha}_1 + z_3(f_3 + b_5u_q),\end{aligned} \tag{2.10}$$

where

$$\begin{aligned}\dot{\alpha}_1 &= -k_1(x_2 - \dot{x}_{1d}) + \ddot{x}_{1d},\\ \dot{\alpha}_2 &= \sum_{i=1}^{2}\frac{\partial\alpha_2}{\partial x_i}\dot{x}_i + \sum_{i=0}^{2}\frac{\partial\alpha_2}{\partial x_{1d}^{(i)}}x_{1d}^{(i+1)} + \frac{\partial\alpha_2}{\partial\hat{J}}\dot{\hat{J}} + \frac{\partial\alpha_2}{\partial x_4}\dot{x}_4\\ &= \frac{\partial\alpha_2}{\partial x_1}x_2 + \frac{\partial\alpha_2}{\partial x_2}\left(\frac{a_1}{J}x_3x_4 - \frac{T_L}{J}\right)\\ &\quad + \sum_{i=0}^{2}\frac{\partial\alpha_2}{\partial x_{1d}^{(i)}}x_{1d}^{(i+1)} + \frac{\partial\alpha_2}{\partial\hat{J}}\dot{\hat{J}} + \frac{\partial\alpha_2}{\partial x_4}(c_1x_4 + b_4x_5),\\ f_3(Z) &= a_1z_2x_4 + b_1x_3 + b_2x_2x_4 - b_3x_2x_5 - b_4\frac{x_3x_5}{x_4} - \dot{\alpha}_2,\\ Z &= [x_1, x_2, x_3, x_4, x_5, x_{1d}, \dot{x}_{1d}, \ddot{x}_{1d}, \hat{J}]^T.\end{aligned} \tag{2.11}$$

Notice that f_3 contains the derivative of α_2, and the unknown parameter J appears in the expression of f_3. This will make the classical adaptive backstepping method become complicated, troubled, and the designed control law u_q will have a complex structure. To avoid this trouble and simplify the control signal structure, we will use the FLS to approximate the nonlinear function f_3. As shown later, the design procedure of u_q becomes simple and u_q is of a simple structure.

According to Lemma 2.1, for any given $\varepsilon_3 > 0$, there exists an FLS $W_3^T S(Z)$ such that

$$f_3(Z) = W_3^T S(Z) + \delta_3(Z), \tag{2.12}$$

where $\delta_3(Z)$ is the approximation error and satisfies $|\delta_3| \le \varepsilon_3$. Consequently, a straightforward calculation deduces the following inequality:

$$z_3f_3(Z) = z_3\left(W_3^T S(Z) + \delta_3(Z)\right) \le \frac{1}{2l_3^2}z_3^2\,\|W_3\|^2\,S^2 + \frac{1}{2}l_3^2 + \frac{1}{2}z_3^2 + \frac{1}{2}\varepsilon_3^2. \tag{2.13}$$

Therefore, it follows immediately from substituting (2.13) into (2.10) that

$$
\begin{aligned}
\dot{V}_3 \leq & -k_1z_1^2 - k_2z_2^2 + z_2(\hat{J} - J)\dot{\alpha}_1 + \frac{1}{2}\varepsilon_2^2 d^2 + \frac{1}{2l_3^2}z_3^2 \left\| W_3 \right\|^2 S^2 \\
& + \frac{1}{2}l_3^2 + \frac{1}{2}z_3^2 + \frac{1}{2}\varepsilon_3^2 + b_5 z_3 u_q.
\end{aligned}
$$

At this present stage, the control law u_q is devised as

$$
u_q = \frac{1}{b_5}(-k_3z_3 - \frac{1}{2}z_3 - \frac{1}{2l_3^2}z_3\hat{\theta}S^2), \tag{2.14}
$$

where $\hat{\theta}$ is the estimation of the unknown constant θ which will be specified later. Furthermore, using the equality (2.14), it can be certified easily that

$$
\dot{V}_3 \leq -\sum_{i=1}^{3} k_i z_i^2 + z_2(\hat{J} - J)\dot{\alpha}_1 + \frac{1}{2l_3^2}z_3^2(\|W_3\|^2 - \hat{\theta})S^2 + \frac{1}{2}l_3^2 + \frac{1}{2}\varepsilon_2^2 d^2 + \frac{1}{2}\varepsilon_3^2.
$$

Step 4: For the reference signal x_{4d}, define the tracking error variable as $z_4 = x_4 - x_{4d}$. From the fourth differential equation of (2.1), one has $\dot{z}_4 = \dot{x}_4 - \dot{x}_{4d}$.

Choose the Lyapunov function candidate as $V_4 = V_3 + \frac{1}{2}z_4^2$. Then the derivative of V_4 is given by

$$
\begin{aligned}
\dot{V}_4 = & \dot{V}_3 + z_4\dot{z}_4 \\
\leq & -\sum_{i=1}^{3} k_i z_i^2 + \frac{1}{2l_3^2}z_3^2(\|W_3\|^2 - \hat{\theta})S^2 + \frac{1}{2}l_3^2 + \frac{1}{2}\varepsilon_2^2 d^2 + \frac{1}{2}\varepsilon_3^2 \\
& + z_2(\hat{J} - J)\dot{\alpha}_1 + z_4\left(c_1x_4 + b_4x_5 - \dot{x}_{4d}\right).
\end{aligned} \tag{2.15}
$$

Now, choose the virtual control law α_3 as

$$
\alpha_3(x_4, x_{4d}, \dot{x}_{4d}) = \frac{1}{b_4}(-k_4z_4 - c_1x_4 + \dot{x}_{4d}), \tag{2.16}
$$

with $k_4 > 0$ being a design parameter. Define $z_5 = x_5 - \alpha_3$. By using (2.16), (2.15) can be rewritten as

$$
\begin{aligned}
\dot{V}_4 \leq & -\sum_{i=1}^{4} k_i z_i^2 + \frac{1}{2l_3^2}z_3^2(\|W_3\|^2 - \hat{\theta})S^2 + \frac{1}{2}l_3^2 + \frac{1}{2}\varepsilon_3^2 \\
& + \frac{1}{2}\varepsilon_2^2 d^2 + z_2(\hat{J} - J)\dot{\alpha}_1 + z_4z_5.
\end{aligned}
$$

Step 5: At this step, we will construct the control law u_d. To this end, choose the Lyapunov function candidate as $V_5 = V_4 + \frac{1}{2}z_5^2$. Then the derivative of V_5 is given by

$$\begin{aligned}\dot{V}_5 &= \dot{V}_4 + z_5\dot{z}_5\\ &\leq -\sum_{i=1}^{4} k_i z_i^2 + \frac{1}{2l_3^2} z_3^2(\|W_3\|^2 - \hat{\theta})S^2 + \frac{1}{2}l_3^2 + \frac{1}{2}\varepsilon_3^2\\ &\quad + \frac{1}{2}\varepsilon_2^2 d^2 + z_2(\hat{J} - J)\dot{\alpha}_1 + z_5(f_5 + b_5 u_d),\end{aligned} \tag{2.17}$$

where $f_5(Z) = z_4 + b_1 x_5 + d_2 x_4 + b_3 x_2 x_3 + b_4 \frac{x_3^2}{x_4} - \dot{\alpha}_3$. Similarly, by Lemma 2.1 the FLS $W_5^T S(Z)$ is employed to approximate the nonlinear function f_5 such that for given $\varepsilon_5 > 0$,

$$z_5 f_5(Z) \leq \frac{1}{2l_5^2} z_5^2 \|W_5\|^2 S^2 + \frac{1}{2}l_5^2 + \frac{1}{2}z_5^2 + \frac{1}{2}\varepsilon_5^2. \tag{2.18}$$

Substituting (2.18) into (2.17) gives

$$\begin{aligned}\dot{V}_5 &= \dot{V}_4 + z_5\dot{z}_5\\ &\leq -\sum_{i=1}^{5} k_i z_i^2 + \frac{1}{2l_3^2} z_3^2(\|W_3\|^2 - \hat{\theta})S^2 + \frac{1}{2}l_3^2 + \frac{1}{2}\varepsilon_3^2 + \frac{1}{2}l_5^2 + \frac{1}{2}\varepsilon_5^2\\ &\quad + \frac{1}{2l_5^2} z_5^2(\|W_5\|^2 - \hat{\theta})S^2 + \frac{1}{2}\varepsilon_2^2 d^2 + z_2(\hat{J} - J)\dot{\alpha}_1 + z_5 b_5 u_d.\end{aligned}$$

Now choose u_d as

$$u_d = \frac{-1}{b_5}(k_5 z_5 + \frac{1}{2}z_5 + \frac{1}{2l_5^2} z_5 \hat{\theta} S^2), \tag{2.19}$$

and define $\theta = \max\{\|W_3\|^2, \|W_5\|^2\}$. Additionally, using the equality (2.19), it can be verified easily that

$$\begin{aligned}\dot{V}_5 \leq &-\sum_{i=1}^{5} k_i z_i^2 + \frac{1}{2l_3^2} z_3^2(\|W_3\|^2 - \hat{\theta})S^T(Z)S(Z) + \frac{1}{2}l_3^2 + \frac{1}{2}\varepsilon_3^2 + \frac{1}{2}\varepsilon_2^2 d^2\\ &+ z_2(\hat{J} - J)\dot{\alpha}_1 + \frac{1}{2l_5^2} z_5^2(\|W_5\|^2 - \hat{\theta})S^T(Z)S(Z) + \frac{1}{2}l_5^2 + \frac{1}{2}\varepsilon_5^2.\end{aligned} \tag{2.20}$$

Define variables $\tilde{J}$ and $\tilde{\theta}$ as

$$\begin{aligned}\tilde{J} &= \hat{J} - J,\\ \tilde{\theta} &= \hat{\theta} - \theta,\end{aligned} \tag{2.21}$$

and select the Lyapunov function candidate as:

$$V = V_5 + \frac{1}{2r_1}\tilde{J}^2 + \frac{1}{2r_2}\tilde{\theta}^2, \tag{2.22}$$

where $r_i, i = 1, 2$ are positive constant. By differentiating V and taking (2.20)-(2.22) into account, one has

$$\begin{aligned}
\dot{V} \leq & -\sum_{i=1}^{5} k_i z_i^2 + \frac{1}{2l_3^2} z_3^2(\|W_3\|^2 - \hat{\theta}) S^T(Z)S(Z) + \frac{1}{2}l_3^2 + \frac{1}{2}\varepsilon_3^2 + \frac{1}{2}\varepsilon_2^2 d^2 + \frac{1}{2}l_5^2 \\
& + z_2(\hat{J} - J)\dot{\alpha}_1 + \frac{1}{2l_5^2} z_5^2(\|W_5\|^2 - \hat{\theta}) S^T(Z)S(Z) + \frac{1}{2}\varepsilon_5^2 + \frac{1}{r_1}\tilde{J}\dot{\hat{J}} + \frac{1}{r_2}\tilde{\theta}\dot{\hat{\theta}} \\
= & -\sum_{i=1}^{5} k_i z_i^2 + \frac{1}{2}l_3^2 + \frac{1}{2}\varepsilon_3^2 + \frac{1}{2}l_5^2 + \frac{1}{2}\varepsilon_5^2 + \frac{1}{2}\varepsilon_2^2 d^2 + \frac{1}{r_1}\tilde{J}\left(r_1 z_2 \dot{\alpha}_1 + \dot{\hat{J}}\right) \\
& + \frac{1}{r_2}\tilde{\theta}\left[-\frac{r_2}{2l_5^2} z_5^2 S^T(Z)S(Z) - \frac{r_2}{2l_3^2} z_3^2 S^T(Z)S(Z) + \dot{\hat{\theta}}\right].
\end{aligned} \tag{2.23}$$

According to (2.23), the corresponding adaptive laws are chosen as follows:

$$\begin{aligned}
\dot{\hat{J}} &= -r_1 z_2 \dot{\alpha}_1 - m_1 \hat{J}, \\
\dot{\hat{\theta}} &= \frac{r_2}{2l_3^2} z_3^2 S^T(Z)S(Z) + \frac{r_2}{2l_5^2} z_5^2 S^T(Z)S(Z) - m_2\hat{\theta},
\end{aligned} \tag{2.24}$$

where m_i, for $i = 1, 2$, l_3 and l_5 are positive constant.

Remark 2.3 To demonstrate the advantage of the adaptive fuzzy backstepping technique over the traditional backstepping summarized in Sect. 2.4, we compare the controller in Eqs. (2.14) and (2.19) with those described in Eqs. (2.35) and (2.39) corresponding, respectively. It can be seen clearly that the expression of the backstepping controller (2.35) and (2.39) would be much more complicated than that of the new controller (2.35) and (2.39). The number of terms in the expression of (2.14) and (2.19) is much larger. This drawback is called the "explosion of terms" above [16]. ♦

Theorem 2.4 *Consider system (2.1) and reference signal x_{1d}. If the virtual control signals are constructed as in (2.4), (2.8) and (2.16), the adaptive law is designed as in (2.24), we choose the adaptive fuzzy controllers (2.14) and (2.19) such that the resulting tracking errors converge to the originąŕs small neighborhood. Also, all closed-loop signals of the controlled system are bounded.*

Proof To address the stability analysis of the resulting closed-loop system, substitute (2.24) into (2.23) to obtain that

$$\dot{V} \leq -\sum_{i=1}^{5} k_i z_i^2 + \frac{1}{2} l_3^2 + \frac{1}{2} \varepsilon_3^2 + \frac{1}{2} l_5^2 + \frac{1}{2} \varepsilon_5^2 + \frac{1}{2} \varepsilon_2^2 d^2 - \frac{m_1}{r_1} \tilde{J} \hat{J} - \frac{m_2}{r_2} \tilde{\theta} \hat{\theta}. \quad (2.25)$$

For the term $-\tilde{J}\hat{J}$, one has $-\tilde{J}\hat{J} \leq -\tilde{J}(\tilde{J} + J) \leq -\frac{1}{2}\tilde{J}^2 + \frac{1}{2}J^2$. Similarly, $-\tilde{\theta}\hat{\theta} \leq -\frac{1}{2}\tilde{\theta}^2 + \frac{1}{2}\theta^2$ holds. Consequently, by using these inequalities (2.25) can be rewritten in the following form.

$$\begin{aligned} \dot{V} &\leq -\sum_{i=1}^{5} k_i z_i^2 - \frac{m_1}{2r_1} \tilde{J}^2 - \frac{m_2}{2r_2} \tilde{\theta}^2 + \frac{1}{2} l_3^2 + \frac{1}{2} \varepsilon_3^2 \\ &\quad + \frac{1}{2} l_5^2 + \frac{1}{2} \varepsilon_5^2 + \frac{1}{2} \varepsilon_2^2 d^2 + \frac{m_1}{2r_1} J^2 + \frac{m_2}{2r_2} \theta^2 \\ &\leq -a_0 V + b_0, \end{aligned} \quad (2.26)$$

where $a_0 = \min\left\{2k_1, \frac{2k_2}{J}, 2k_3, 2k_4, 2k_5, m_1, m_2\right\}$ and $b_0 = \frac{1}{2}l_3^2 + \frac{1}{2}\varepsilon_3^2 + \frac{1}{2}l_5^2 + \frac{1}{2}\varepsilon_5^2 + \frac{1}{2}\varepsilon_2^2 d^2 + \frac{m_1}{2r_1} J^2 + \frac{m_2}{2r_2}\theta^2$. Furthermore, (2.26) implies that

$$V(t) \leq (V(t_0) - \frac{b_0}{a_0})e^{-a_0(t-t_0)} + \frac{b_0}{a_0} \leq V(t_0) + \frac{b_0}{a_0}, \forall t \geqslant t_0. \quad (2.27)$$

As a result, all $z_i (i = 1, 2, 3, 4, 5)$, $\tilde{J}$ and $\tilde{\theta}$ belong to the compact set

$$\Omega = \left\{ (z_i, \tilde{J}, \tilde{\theta}) | V \leq V(t_0) + \frac{b_0}{a_0}, \forall t \geqslant t_0 \right\}.$$

Namely, all the signals in the closed-loop system are bounded. Especially from (2.27) we have

$$\lim_{t\to\infty} z_1^2 \leq \frac{2b_0}{a_0}.$$

From the definitions of a_0 and b_0, it is clear that to get a small tracking error by taking r_i sufficiently large and l_i and ε_i small enough after giving the parameters k_i and m_i. ■

2.4 Simulation Results

In this section, we will compare the proposed method and the conventional backstepping technique. More concretely, the classical backstepping is first used to control design for the system (2.1), and the simulation is implemented by both the proposed approach and the classical one.

2.4.1 Classical Backstepping Design

This subsection devotes to design controllers by classical backstepping approach.

Step 1: For the reference signal x_{1d}, define the tracking error variable as $z_1 = x_1 - x_{1d}$. From the first differential equation of (2.1), the error dynamic system is given by $\dot{z}_1 = x_2 - \dot{x}_{1d}$.

Choose Lyapunov function candidate as $V_1 = \frac{1}{2}z_1^2$, then the time derivative of V_1 is computed by

$$\dot{V}_1 = z_1\dot{z}_1 = z_1(x_2 - \dot{x}_{1d}). \tag{2.28}$$

Construct the virtual control law α_1 as

$$\alpha_1 = -k_1 z_1 + \dot{x}_{1d}, \tag{2.29}$$

with $k_1 > 0$ being a design parameter and $z_2 = x_2 - \alpha_1$. By using (2.29), (2.28) can be rewritten of the following form.

$$\dot{V}_1 = -k_1 z_1^2 + z_1 z_2.$$

Step 2: Differentiating z_2 gives

$$\dot{z}_2 = \dot{x}_2 - \dot{\alpha}_1 = \frac{a_1}{J}x_3x_4 - \frac{T_L}{J} - \dot{\alpha}_1. \tag{2.30}$$

Now, choose the Lyapunov function candidate as $V_2 = V_1 + \frac{J}{2}z_2^2$. Obviously, the time derivative of V_2 is given by

$$\dot{V}_2 = \dot{V}_1 + \frac{J}{2}z_2\dot{z}_2 = -k_1 z_1^2 + z_2(z_1 + a_1x_3x_4 - T_L - J\dot{\alpha}_1). \tag{2.31}$$

The virtual control α_2 is constructed as

$$\alpha_2 = \frac{1}{a_1x_4}(-k_2z_2 - z_1 + T_L + J\dot{\alpha}_1), \tag{2.32}$$

where $k_2 > 0$ is a positive design parameter and $\dot{\alpha}_1 = -k_1(x_2 - \dot{x}_{1d}) + \ddot{x}_{1d}$. Adding and subtracting α_2 in the bracket in (2.31) shows that

$$\dot{V}_2 = -k_1z_1^2 - k_2z_2^2 + a_1z_2z_3x_4, \tag{2.33}$$

with $z_3 = x_3 - \alpha_2$.

Step 3: Differentiating z_3 results in the following differential equation.

$$\dot{z}_3 = \dot{x}_3 - \dot{\alpha}_2 = b_1x_3 + b_2x_2x_4 - b_3x_2x_5 - b_4\frac{x_3x_5}{x_4} + b_5u_q - \dot{\alpha}_2.$$

Choose the Lyapunov function candidate as $V_3 = V_2 + \frac{1}{2}z_3^2$. Then, differentiating V_3 holds

$$\dot{V}_3 = \dot{V}_2 + z_3\dot{z}_3 = -k_1z_1^2 - k_2z_2^2 + z_3(a_1z_2x_4 + b_1x_3 + b_2x_2x_4 - b_3x_2x_5 - b_4\frac{x_3x_5}{x_4} + b_5u_q - \dot{\alpha}_2), \tag{2.34}$$

where

$$\begin{aligned}\dot{\alpha}_2 &= \sum_{i=1}^{2}\frac{\partial\alpha_2}{\partial x_i}\dot{x}_i + \sum_{i=0}^{2}\frac{\partial\alpha_2}{\partial x_{1d}^{(i)}}x_{1d}^{(i+1)} + \frac{\partial\alpha_2}{\partial x_4}\dot{x}_4\\ &= \frac{\partial\alpha_2}{\partial x_1}x_2 + \frac{\partial\alpha_2}{\partial x_2}\left(\frac{a_1}{J}x_3x_4 - \frac{T_L}{J}\right) + \sum_{i=0}^{2}\frac{\partial\alpha_2}{\partial x_{1d}^{(i)}}x_{1d}^{(i+1)} + \frac{\partial\alpha_2}{\partial x_4}(c_1x_4 + b_4x_5).\end{aligned}$$

And the control law u_q is designed as

$$\begin{aligned}u_q = &-\frac{1}{b_5}(k_3z_3 + a_1z_2x_4 + b_1x_3 + b_2x_2x_4 - b_3x_2x_5 - b_4\frac{x_3x_5}{x_4})\\ &+\frac{1}{b_5}\left(\frac{\partial\alpha_2}{\partial x_1}x_2 + \frac{\partial\alpha_2}{\partial x_2}\left(\frac{a_1}{J}x_3x_4 - \frac{T_L}{J}\right) + \frac{\partial\alpha_2}{\partial x_4}(c_1x_4 + b_4x_5)\right.\\ &\left.+\sum_{i=0}^{2}\frac{\partial\alpha_2}{\partial x_{1d}^{(i)}}x_{1d}^{(i+1)}\right).\end{aligned} \tag{2.35}$$

Furthermore, using the equality (2.35), it can be certified easily that

$$\dot{V}_3 \le -\sum_{i=1}^{3}k_iz_i^2.$$

Step 4: For the reference signal x_{4d}, define the tracking error variable as $z_4 = x_4 - x_{4d}$. From the fourth differential equation of (2.1), one has $\dot{z}_4 = \dot{x}_4 - \dot{x}_{4d}$. Choose the Lyapunov function candidate as $V_4 = V_3 + \frac{1}{2}z_4^2$. Then the derivative of V_4 is given by

$$\dot{V}_4 = \dot{V}_3 + z_4\dot{z}_4 \le -\sum_{i=1}^{3}k_iz_i^2 + z_4\left(c_1x_4 + b_4x_5 - \dot{x}_{4d}\right). \tag{2.36}$$

Now, constitute the virtual control law α_3 as

$$\alpha_3 = \frac{1}{b_4}(-k_4z_4 - c_1x_4 + \dot{x}_{4d}), \tag{2.37}$$

with $k_4 > 0$ being a design parameter. Define $z_5 = x_5 - \alpha_3$. By using (2.37), (2.36) can be expressed as

$$\dot{V}_4 \le -\sum_{i=1}^{4} k_i z_i^2 + z_4 z_5.$$

Step 5: At this step, we will construct the control law u_d. To this end, choose the Lyapunov function candidate as $V_5 = V_4 + \frac{1}{2}z_5^2$. Then the derivative of V_5 is given by

$$\begin{aligned}\dot{V}_5 &= \dot{V}_4 + z_5\dot{z}_5\\ &\le -\sum_{i=1}^{4} k_i z_i^2 + z_5(z_4 + b_1x_5 + d_2x_4 + b_3x_2x_3 + b_4\frac{x_3^2}{x_4} - \dot{\alpha}_3 + b_5u_d).\end{aligned} \tag{2.38}$$

Now design u_d as

$$\begin{aligned}u_d &= \frac{-1}{b_5}(k_5z_5 + z_4 + b_1x_5 + d_2x_4 + b_3x_2x_3 + b_4\frac{x_3^2}{x_4} - \dot{\alpha}_3)\\ &= \frac{-1}{b_5}(k_5z_5 + z_4 + b_1x_5 + d_2x_4 + b_3x_2x_3 + b_4\frac{x_3^2}{x_4})\\ &\quad + \frac{1}{b_4b_5}\left[(-k_4 - c_1)\dot{x}_4 + k_4\dot{x}_{4d} + \ddot{x}_{4d})\right],\end{aligned} \tag{2.39}$$

with $k_5 > 0$.

So far, by comparing the controllers (2.14) and (2.19) with the controllers (2.35) and (2.39), it is easy to see that the proposed adaptive fuzzy controllers have a much more simple structure than the classical ones. This means that the proposed adaptive fuzzy controllers are easy to be carried out in practical engineering. Furthermore, the controllers (2.14) and (2.19) are constructed under the assumption that the nonlinear system dynamics functions are unknown. Therefore, the developed control strategy can be used to control the IM system. Since the controllers (2.35) and (2.39) require accurate information on the nonlinear functions, theoretically, when the functions are unknown the classical backstepping cannot be used to construct the controllers (2.35) and (2.39).

2.4.2 Simulation

In order to prove the feasibility of scheme, the proposed adaptive fuzzy controllers (2.14) and (2.19) and the classical backstepping controllers (2.35) and (2.39) will be employed to control the following IM system, respectively. The simulation is implemented for IM with the parameters: $J = 0.0586\,\text{kgm}^2$, $R_s = 0.1$, $\mathrm{R_r} = 0.15$, $L_s = L_r = 0.0699\text{H}$, $L_m = 0.068\text{H}$, $n_p = 1$.

The desired signals are taken as $x_{1d} = 0.5\sin(t) + 0.5\sin(0.5t)$ and $x_{4d} = 1$ with T_L being

$$T_L = \begin{cases} 1.5, & 0 \le t \le 5, \\ 3, & t \ge 5. \end{cases}$$

The proposed adaptive fuzzy controllers are used to control this IM. The control parameters are selected as follows:

$$k_1 = 200, k_2 = 80, = 300, k_4 = k_5 = 100, r_1 = r_2 = r_3 = r_4 = 0.05,$$
$$m_1 = m_2 = 0.05, l_3 = l_4 = 0.5.$$

The fuzzy membership functions are:

$$\mu_{F_i^1} = \exp\left[\frac{-(x+5)^2}{2}\right], \quad \mu_{F_i^2} = \exp\left[\frac{-(x+4)^2}{2}\right],$$
$$\mu_{F_i^3} = \exp\left[\frac{-(x+3)^2}{2}\right], \quad \mu_{F_i^4} = \exp\left[\frac{-(x+2)^2}{2}\right],$$
$$\mu_{F_i^5} = \exp\left[\frac{-(x+1)^2}{2}\right], \quad \mu_{F_i^6} = \exp\left[\frac{-(x-0)^2}{2}\right],$$
$$\mu_{F_i^7} = \exp\left[\frac{-(x-1)^2}{2}\right], \quad \mu_{F_i^8} = \exp\left[\frac{-(x-2)^2}{2}\right],$$
$$\mu_{F_i^9} = \exp\left[\frac{-(x-3)^2}{2}\right], \quad \mu_{F_i^{10}} = \exp\left[\frac{-(x-4)^2}{2}\right],$$
$$\mu_{F_i^{11}} = \exp\left[\frac{-(x-5)^2}{2}\right].$$

The simulation for adaptive fuzzy control is implemented under the assumption that the system parameters and the nonlinear functions are unknown.

Then, the controllers (2.35) and (2.39) are also utilized to control the systems. The corresponding controller parameters are taken as

$$k_1 = 100, \quad k_2 = 50, \quad k_3 = 60, \quad k_4 = 80, \quad k_5 = 20.$$

The simulation for classical backstepping control is carried out assuming that the system parameters and the nonlinear functions are all known.

The simulation results for both cases of adaptive fuzzy control and classical backstepping control are shown by Figs. 2.1, 2.2, 2.3, 2.4, 2.5, 2.6, 2.7 and 2.8. Figures 2.1, 2.2, 2.3 and 2.4 display the system output responses and the reference signals for both control approaches, Figs. 2.5, 2.6, 2.7 and 2.8 show the control input signals. From Figs. 2.1, 2.2, 2.3 and 2.4, it is seen clearly that under the actions of controllers

(2.14) and (2.19) and the controllers (2.35) and (2.39), the system outputs follow the desired reference signals well.

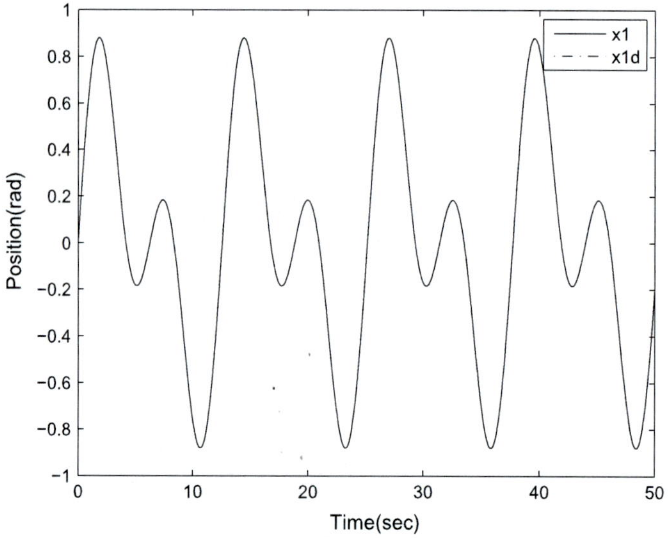

Fig. 2.1 Trajectories of the x_1 and x_{1d}

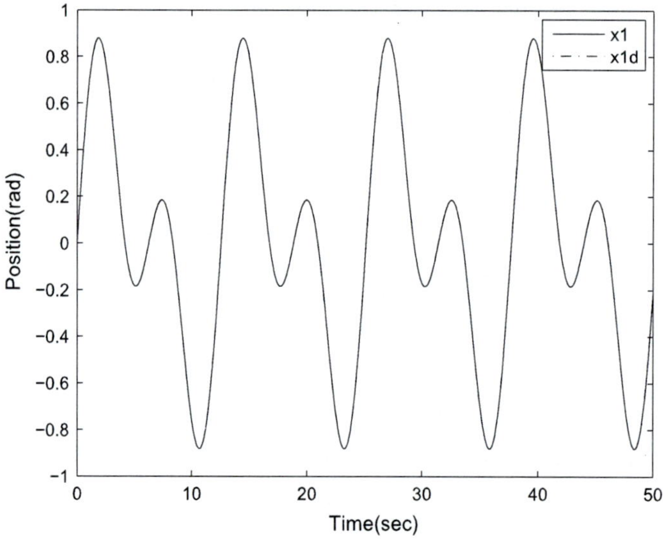

Fig. 2.2 Trajectories of the x_1 and x_{1d} for classical backstepping

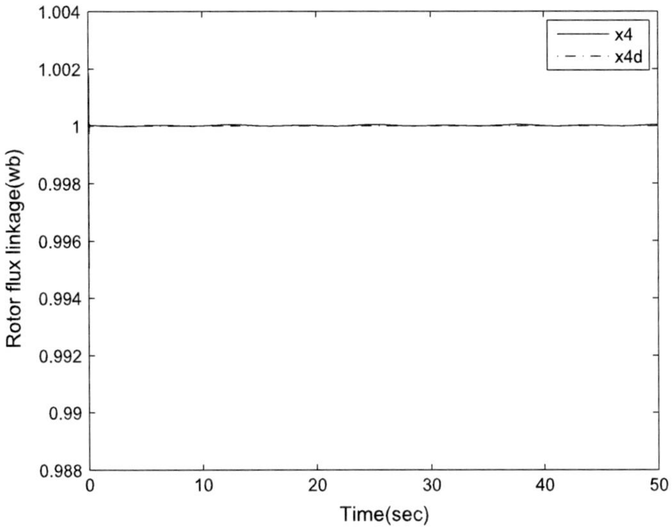

Fig. 2.3 Trajectories of the x_4 and x_{4d}

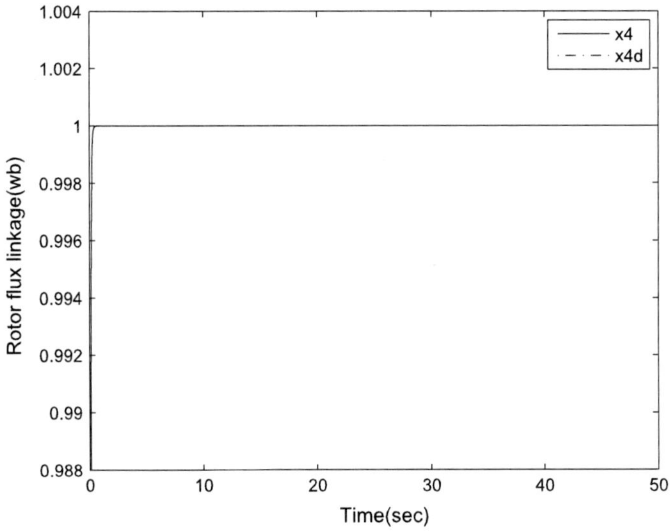

Fig. 2.4 Trajectories of the x_4 and x_{4d} for classical backstepping

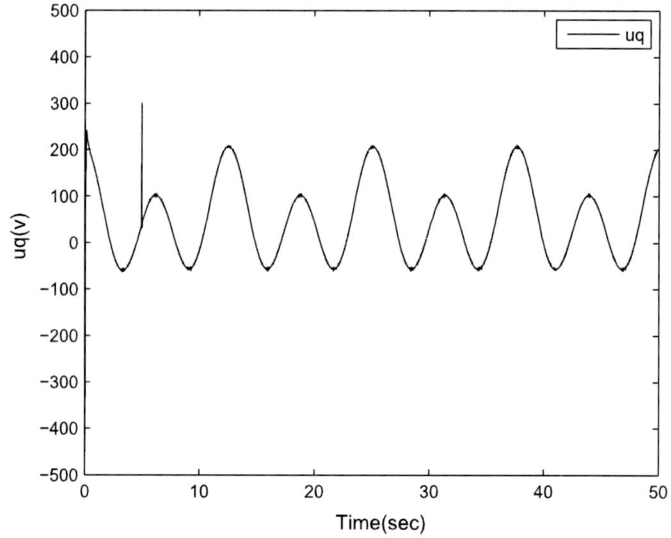

Fig. 2.5 Curve of the u_q

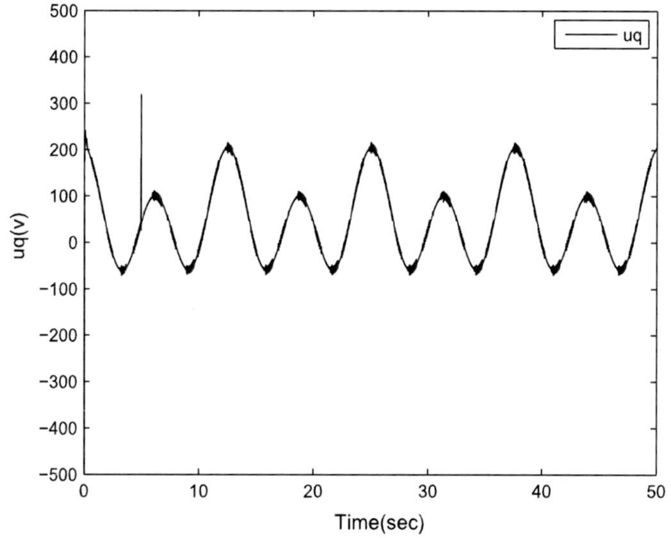

Fig. 2.6 Curve of the u_q for classical backstepping

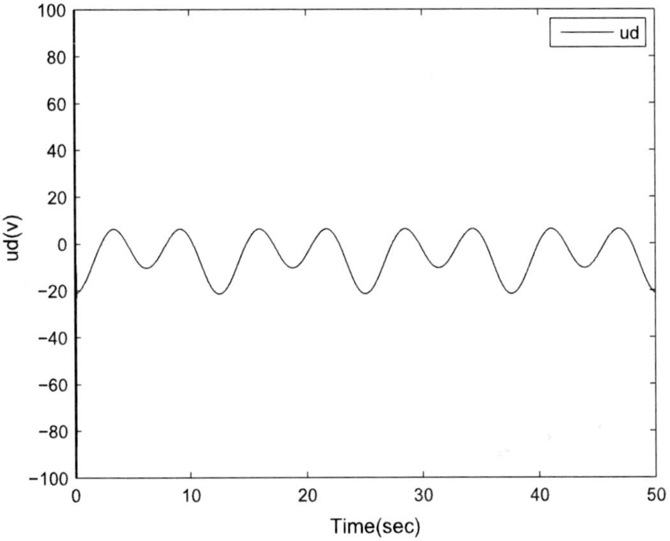

Fig. 2.7 Curve of the u_d

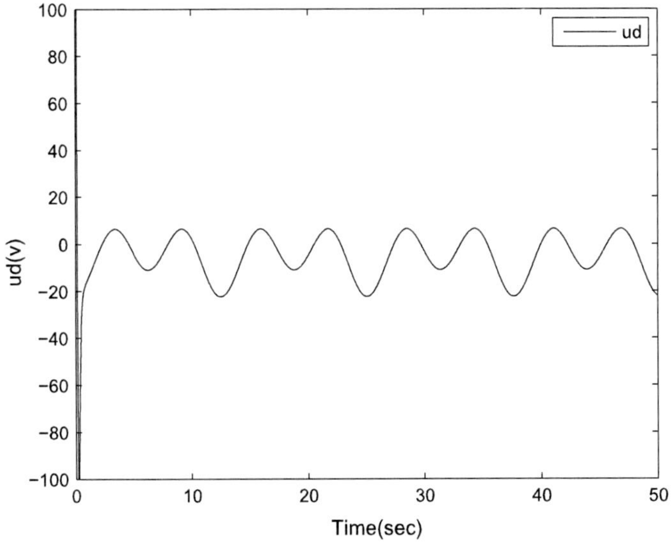

Fig. 2.8 Curve of the u_d for classical backstepping

2.5 Conclusion

In this chapter an adaptive fuzzy control scheme is proposed to control IM. The proposed controllers that overcome the traditional backstepping's major problems guarantee that the tracking error converges to a small neighborhood of the origin and all the closed-loop signals are bounded. It is demonstrated that the proposed control method ensures desired tracking and the boundedness of all signals with the parameter uncertainties and load disturbances.

References

1. Yazdanpanah, R., Soltani, J., Markadeh, G.R.A: Nonlinear torque and stator flux controller for induction motor drive based on adaptive input-output feedback linearization and sliding mode control. Energy Conv. Manag. **49**(4), 541–550 (2008)
2. Haddoun, A., Benbouzid, M.E.H., Diallo, D., Abdessemed, R., Ghouili, J., Srairi, K.: A loss-minimization DTC scheme for EV induction motors. IEEE Trans. Veh. Technol. **56**, 81–88 (2007)
3. Veselic, B., Perunicic-Drazenovicm, B., Milosavljevic, C.S.: High-performance position control of induction motor using discrete-time sliding-mode control. IEEE Trans. Ind. Electron. **55**(11), 3809–3817 (2008)
4. Hazzab, A., Bousserhane, I.K., Zerbo, M., Sicard, P.: Real time implementation of fuzzy gain scheduling of PI controller for induction motor machine control. Neural Process. Lett. **24**(3), 203–215 (2006)
5. Ponmani, C.: Performance improvement of matrix converter fed induction motor under input voltage and load disturbances using internal model control. Int. J. Electr. Power Energy Syst. **44**(1), 43–51 (2013)
6. Traore, D., Leon, J.D., Glumineau, A.: Sensorless induction motor adaptive observer-backstepping controller: experimental robustness tests on low frequencies benchmark. IET Contr. Theory Appl. **4**(10), 1989–2002 (2001)
7. Shieh, H.J., Shyu, K.K.: Nonlinear sliding-mode torque control with adaptive backstepping approach for induction motor drive. IEEE Trans. Ind. Electron. **46**(2), 380–389 (1999)
8. Zaafouri, A., Regaya, C.B., Azza, H.B.: DSP-based adaptive backstepping using the tracking errors for high-performance sensorless speed control of induction motor drive. ISA Trans. **60**, 333–347 (2016)
9. Yu, J.P., Ma, Y.M., Yu, H.S., Lin, C.: Adaptive fuzzy dynamic surface control for induction motors with iron losses in electric vehicle drive systems via backstepping. Inf. Sci. **376**, 172–189 (2017)
10. Chen, B., Liu, X., Tong, S.: Adaptive fuzzy approach to control unified chaotic systems. Chaos Solitons Fractals **34**, 1180–1187 (2007)
11. Tu, K.Y., Lee, T.T., Wang, W.J.: Design of a multi-layer fuzzy logic controller for multi-input multi-output systems. Fuzzy Sets Syst. **111**(2), 199–214 (2000)
12. Wai, R.J., Lin, K.M., Lin, C.Y.: Total sliding-mode speed control of field oriented induction motor servo drive. In: Proceedings of the 5th Asian Control Conference, Melbourne, Australia, pp. 1354–1361 (2004)
13. Liu, X.P., Gu, G.X., Zhou, K.M.: Robust stabilization of MIMO nonlinear systems by backstepping. Automatica **35**, 987–992 (1999)
14. Li, K., Zhang, C.H., Cui, N.X.: Vector control of induction motor for electric vehicles considering iron losses and its energy optimization strategy. Control Theory Appl. **24**, 959–963 (2007)

15. Marino, R., Peresada, S., Valigi, P.: Adaptive input-output linearizing control of induction motors. IEEE Trans. Autom. Control **38**(2), 208–221 (1993)
16. Wang, L.X., Mendel, J.M.: Fuzzy basis functions, universal approximation, and orthogonal least squares learning. IEEE Trans. Neural Netw. **3**(5), 807–814 (1992)

Chapter 3
NNs-Based Command Filtered Control for IM with Input Saturation

In this chapter, the neural networks approximation-based command filtered adaptive control is proposed for induction motor (IM) with input saturation. The neural networks are used to approximate the nonlinearities, and the command filtered method is used to deal with the "explosion of complexity" problem caused by the derivative of virtual controllers in the traditional backstepping design. The compensating signals are further exploited to get rid of the drawback caused by the dynamics surface technology. It is verified that the adaptive neural controller guarantees that the tracking error can converge to a small neighborhood of the origin. At last, the effectiveness and advantages of the proposed method are illustrated by simulation results.

3.1 Introduction

The variable voltage and frequency of IM are usually utilized to control the speed and torque of the IM [1]. Voltage/Frequency (V/F) scalar control strategy is applicable for IM to develop the performance and dynamic response of the IM, and it has a few advantages such as simple structure, low cost, easy design, and low steady-state error [2]. The scalar control strategy of an IM is simple to carry out and offer a good satisfactory steady-state response, but poor in dynamic response. Considering that the dynamic model of IM is highly nonlinear and multivariable, the task is still hot and difficult to obtain excellent control property. The researchers have developed many nonlinear control methods such as input-output linearization control [3], sliding mode control [4–6], backstepping control to achieve high performance control for the IM [7–10].

In another research front line, the adaptive control methods via approximation theories have been introduced to dispose of the nonlinear systems with parametric

J. Yu et al., *Intelligent Backstepping Control for the Alternating-Current Drive Systems*, Studies in Systems, Decision and Control 349,
https://doi.org/10.1007/978-3-030-67723-7_3

uncertainty based on fuzzy logic system (FLS) or neural networks (NNs) approximation [11]. Adaptive fuzzy/NNs backstepping gives a system methodology for coping with the nonlinearities in the adaptive backstepping method. Additionally, the command filtered control (CFC) method is used to solve the "explosion of complexity" problem by introducing certain command signals and their derivatives, which may eliminate the requirement of analytic differentiation. Meanwhile, the error compensation mechanism is put forward to reduce the filtered errors caused by the command filters. Thus, the tracking performance with CFC scheme will be better than the DSC technology. In addition, the actuator is constrained in most practical engineering systems, so the input saturation should be considered for IM, but only a few researchers consider it in practical engineering systems [12–14].

In this chapter, the adaptive NNs approximation-based CFC method is proposed for IM systems with input saturation. By utilizing the output of a command filter to approximate the derivative of the virtual control at each step of backstepping, the problem of "explosion of complexity" can be eradicated. And the errors caused by command filters can be reduced by introducing the compensation signals. The proposed method can guarantee that all signals in the closed-loop system are bounded despite of the existence of input saturation.

3.2 Mathematical Model of IM Drive System

From (1.11), the dynamic model of IM drivers can be described by the following differential equations:

$$\begin{cases} \dot{x}_1 = \frac{a_1}{J} x_2 x_3 - \frac{T_L}{J} \\ \dot{x}_2 = b_1 x_2 + b_2 x_1 x_3 - b_3 x_1 x_4 - b_4 \frac{x_2 x_4}{x_3} + b_5 u_q \\ \dot{x}_3 = c_1 x_3 + b_4 x_4 \\ \dot{x}_4 = b_1 x_4 + d_2 x_3 + b_3 x_1 x_2 + b_4 \frac{x_2^2}{x_3} + b_5 u_d \end{cases} \tag{3.1}$$

where x_1 and x_3 denote the rotor position and rotor flux linkage. i_d and i_q stand for the $d - q$ axis currents. u_q and u_d are the scalar control signals. If we assume u is also the scalar control signal and it denotes the plant input subject to nonsymmetric saturated nonlinearity described by

$$u = sat(\vartheta) = \begin{cases} u_{\max}, \vartheta \geq u_{\max} \\ \vartheta, u_{\min} < \vartheta < u_{\max} \\ u_{\min}, \vartheta \leq u_{\min} \end{cases}$$

where $u_{\max} > 0$ and $u_{\min} < 0$ are unknown constants of input saturation and ϑ is the input signal of the saturated nonlinearity.

$$g(\vartheta) = \begin{cases} u_{\max} * \tanh\left(\frac{\vartheta}{u_{\max}}\right), \vartheta \geq 0 \\ u_{\min} * \tanh\left(\frac{\vartheta}{u_{\min}}\right), \vartheta < 0 \end{cases}$$

$$= \begin{cases} u_{\max} * \frac{e^{\frac{\vartheta}{u_{\max}}} - e^{-\frac{\vartheta}{u_{\max}}}}{e^{\frac{v}{u_{\max}}} + e^{-\frac{v}{u_{\max}}}}, \vartheta \geq 0 \\ u_{\min} * \frac{e^{\frac{\vartheta}{u_{\min}}} - e^{-\frac{\vartheta}{u_{\min}}}}{e^{\frac{\vartheta}{u_{\min}}} + e^{-\frac{\vartheta}{u_{\min}}}}, \vartheta < 0 \end{cases}$$

$u = sat(\vartheta) = g(\vartheta) + d(\vartheta) \Longrightarrow |d(\vartheta)| = |sat(\vartheta) - g(\vartheta)| \leq \max \{u_{\max}(1 - \tanh(1)), u_{\min}(\tanh(1) - 1)\} = D$.

In addition, the mean-value theorem implies that there exists a constant $\mu(0 < \mu < 1)$, such that

$$g(\vartheta) = g(\vartheta_0) + g_{\vartheta_\mu}(\vartheta - \vartheta_0),$$

where $g_{\vartheta_\mu} = \left(\frac{\partial g(\vartheta)}{\partial \vartheta}\right)|_{\vartheta=\vartheta_\mu}$ and $\vartheta_\mu = \mu \cdot \vartheta + (1 - \mu)\vartheta_0$.

By choosing $\vartheta_0 = 0$, the above function can be rewritten as $g(\vartheta) = g_{\vartheta_\mu} \cdot \vartheta$, so it has

$$u = g_{\vartheta_\mu} \cdot \vartheta + d(\vartheta).$$

In addition, there exists a positive number g_m, such that $0 < g_m < g_{\vartheta_\mu} \leq 1$. Similarly, the u_q and u_d in the IM can be rewritten as

$$u_q = g_{\vartheta_{\mu q}} \cdot \vartheta_q + d(\vartheta_q); u_d = g_{\vartheta_{\mu d}} \cdot \vartheta_d + d(\vartheta_d).$$

It has been proven in [15] that, for given scalar $\varepsilon > 0$, by choosing sufficiently large l, the RBF NNs can approximate any continuous function over a compact set $\Omega_z \in R^q$ to arbitrary accuracy as $\varphi(z) = \phi^T P(z) + \delta(z)\ \forall z \in \Omega_z \subset R^q$ where $\delta(z)$ is the approximation error, satisfying $|\delta(z)| \leq \varepsilon$ and ϕ is an unknown ideal constant weight vector, which is an artificial quantity required for analytical purpose. Typically, ϕ is chosen as the value of ϕ^* that minimizes $|\delta(z)|$ for all $z \in \Omega_z$,i.e.,

$$\phi := \arg \min_{\phi^* \in R^n} \left\{ \sup_{z \in \Omega_z} \left|\varphi(z) - \phi^{*T} P(z)\right| \right\}.$$

Lemma 3.1 *The form of command-filtered is as below*

$$\dot{\varphi}_1 = \omega_n \varphi_2,$$
$$\dot{\varphi}_2 = -2\zeta\omega_n\varphi_2 - \omega_n(\varphi_1 - \alpha_1),$$

if the input signal α_1 satisfies $|\dot{\alpha}_1| \leq \rho_1$ and $|\ddot{\alpha}_1| \leq \rho_2$ for all $t \geq 0$, where ρ_1 and ρ_2 are positive constants and $\varphi_1(0) = \alpha_1(0)$, $\varphi_2(0) = 0$, then for any $\mu > 0$, there exist $\omega_n > 0$ and $\zeta \in (0, 1]$, such that $|\varphi_1 - \alpha_1| \leq \mu$, $|\dot{z}_1|$, $|\ddot{z}_1|$ and $|\dddot{z}_1|$ are bounded.

Remark 3.2 The ω_n is the command filter natural frequency. By increasing the ω_n, we can decrease the $|\varphi_1 - \alpha_1|$ by decreasing μ. ♦

3.3 Command-Filtered Adaptive NNs Control Design

This section will design an adaptive NNs approximation-based command filtered control method. The errors caused by the command filtered will bring difficulties to get the desired tracking effect. So, the compensation mechanism is developed to compensate the errors, and the error compensation signal ξ_i will be defined later. In addition, Fig. 3.1 shows the signal flow in block diagram form.

Step 1: Define tracking error variable $z_1 = x_1 - x_{1d}$, where the x_{1d} is the given reference signal. Design the compensated error signal as $\nu_1 = z_1 - \xi_1$. Consider the Lyapunov function candidate:

$$V_1 = \frac{J}{2}\nu_1^2.$$

Then we have

$$\dot{V}_1 = J\nu_1\dot{\nu}_1 = \nu_1(a_1 x_2 x_3 - T_L - J\dot{x}_{1d} - J\dot{\xi}_1). \tag{3.2}$$

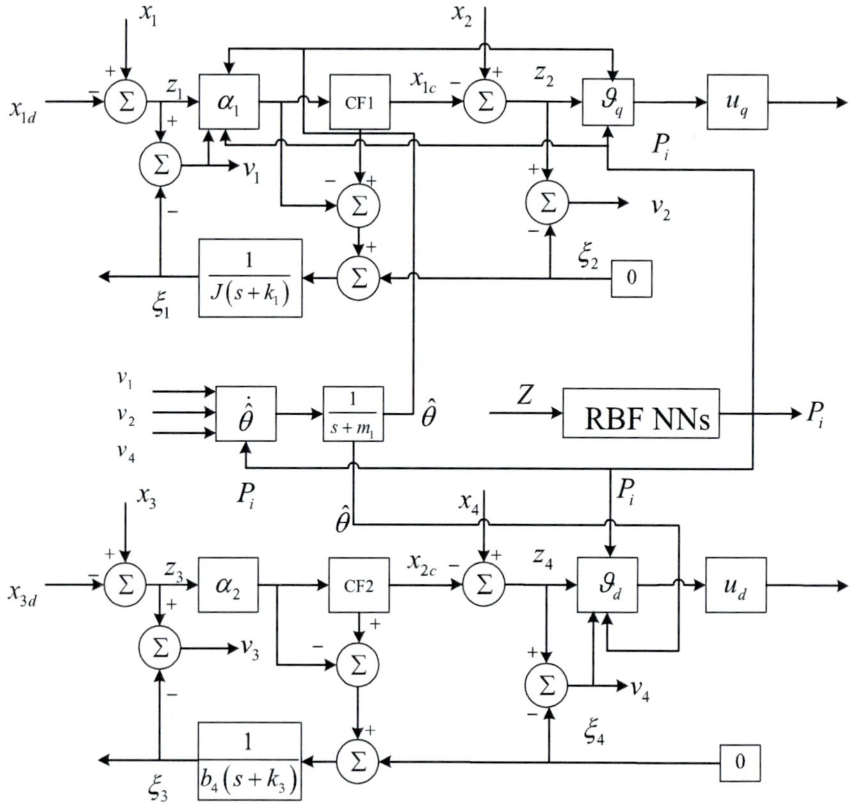

Fig. 3.1 Block diagram of the CFC method for IM

For the actual IM, the parameter T_L is bounded in practice systems and its upper limit is assumed to be $d > 0$, that is, $|T_L| \leq d$. Obviously,

$$-\nu_1 T_L \leq \frac{1}{2\varepsilon_5^2}\nu_1^2 + \frac{1}{2}\varepsilon_5^2 d^2,$$

where ε_5 is an arbitrary small positive constant. Let

$$f_1(Z) = a_1 x_2 x_3 + \frac{1}{2\varepsilon_5^2}\nu_1 - x_2,$$

and $Z = [x_1, x_2, x_3, x_4, x_{1d}, \dot{x}_{1d}]$. For given $\varepsilon_1 > 0$, there exists a RBF NNs $\phi_1^T P_1(Z)$ such that $f_1(Z) = \phi_1^T P_1(Z) + \delta_1(Z)$, where $\delta_1(Z)$ is the approximation error and satisfies $|\delta_1| \leq \varepsilon_1$.

Using the Young's inequality, it has:

$$\begin{aligned}\nu_1 f_1(Z) &= \nu_1 \left(\phi_1^T P_1(Z) + \delta_1(Z)\right)\\ &\leq \frac{1}{2l_1^2}\nu_1^2 \|\phi_1\|^2 P_1^T(Z) P_1(Z) + \frac{1}{2}l_1^2 + \frac{1}{2}\nu_1^2 + \frac{1}{2}\varepsilon_1^2.\end{aligned} \tag{3.3}$$

Selecting the virtual control law α_1 and the compensating signal ξ_1 as

$$\begin{aligned}\alpha_1 &= -k_1 z_1 - \frac{1}{2}\nu_1 - \frac{1}{2l_1^2}\nu_1 \hat{\theta} P_1^T P_1 + \hat{J}\dot{x}_{1d},\\ \dot{\xi}_1 &= \frac{1}{J}\left[-k_1\xi_1 + \xi_2 + (x_{1,c} - \alpha_1)\right],\end{aligned}$$

where the control gain $k_1 > 0$, $\xi_1(0) = 0$, $\hat{\theta}$ will be defined later and $x_{1,c}$ is the output of the command filtered with the input signal α_1. Similarly, we define the compensated error signal $\nu_2 = z_2 - \xi_2$, $z_2 = x_2 - x_{1,c}$. It can be obtained that $\|\xi_1\|$ is bounded by invoking Lemma 3.1 in [16]. If $t \to \infty$, it has

$$\lim_{t\to\infty} \|\xi_1\| \leq \frac{\mu\rho}{2c_0},$$

where $c_0 = \frac{1}{2}\min(k_i)$ with $i = 1, 2, 3, 4$ and $\rho > \max\{1, b_4, b_5\}$. Substituting $\dot{\xi}_1$ and (3.3) into (3.2), it has

$$\begin{aligned}\dot{V}_1 \leq& -k_1\nu_1^2 + \frac{1}{2}\varepsilon_5^2 d^2 + \frac{1}{2l_1^2}\nu_1^2(\|\phi_1\|^2 - \hat{\theta})P_1^T P_1\\ &+ \frac{1}{2}l_1^2 + \frac{1}{2}\varepsilon_1^2 + \nu_1\nu_2 + \nu_1(\hat{J} - J)\dot{x}_{1d}.\end{aligned} \tag{3.4}$$

Step 2: Differentiating z_2 gives

$$\dot{z}_2 = \dot{x}_2 - \dot{x}_{1,c} = b_1 x_2 + b_2 x_1 x_3 - b_3 x_1 x_4 - b_4 \frac{x_2 x_4}{x_3} + b_5 u_q - \dot{x}_{1,c}.$$

Design the compensated tracking signal as $\nu_2 = z_2 - \xi_2$. Consider the following Lyapunov function candidate:

$$V_2 = V_1 + \frac{1}{2}\nu_2^2.$$

Its time derivative is

$$\begin{aligned}\dot{V}_2 \leq & -k_1\nu_1^2 + \frac{1}{2}\varepsilon_5^2 d^2 + \frac{1}{2l_1^2}\nu_1^2(\|\phi_1\|^2 - \hat{\theta})P_1^T P_1 \\ & + \frac{1}{2}l_1^2 + \frac{1}{2}\varepsilon_1^2 + \nu_1(\hat{J} - J)\dot{x}_{1d} + \nu_2(f_2 + b_5 u_q - \dot{\xi}_2), \end{aligned} \tag{3.5}$$

where

$$\begin{aligned} f_2(Z) &= b_1 x_2 + b_2 x_1 x_3 - b_3 x_1 x_4 - b_4 \frac{x_2 x_4}{x_3} + \nu_1 - \dot{x}_{1,c} \\ &= \phi_2^T P_2(Z) + \delta_2(Z). \end{aligned}$$

Similarly, we also can get

$$\nu_2 f_2(Z) \leq \frac{1}{2l_2^2}\nu_2^2 \|\phi_2\|^2 P_2^T(Z) P_2(Z) + \frac{1}{2}l_2^2 + \frac{1}{2}\nu_2^2 + \frac{1}{2}\varepsilon_2^2. \tag{3.6}$$

Let

$$u_q = g_{\vartheta_{\mu q}} \vartheta_q + d\left(\vartheta_q\right),$$

with $\dot{\xi}_2 = 0$, and ϑ_q is designed as

$$\vartheta_q = -k_2 z_2 - \frac{1}{2l_2^2}\nu_2 \hat{\theta} P_2^T P_2. \tag{3.7}$$

Then, we have

$$b_5 u_q \nu_2 = -k_2 b_5 g_{\vartheta_{\mu q}} \nu_2^2 - \frac{1}{2l_2^2}\nu_2^2 b_5 g_{\vartheta_{\mu q}} \hat{\theta} P_2^T P_2 + \nu_2 b_5 d\left(\vartheta_q\right). \tag{3.8}$$

From $0 < g_{mq} < g_{\vartheta_{\mu q}} \leq 1$, there exists a positive number b_q that $b_5 g_{\vartheta_{\mu q}} \geq b_q$. By substituting $\nu_2 b_5 d\left(\vartheta_q\right) \leq \frac{1}{2}\nu_2^2 + \frac{1}{2}b_5^2 D_q^2$, it has

$$b_5 u_q \nu_2 \leq -k_2 b_5 g_{\vartheta_{\mu q}} \nu_2^2 - \frac{1}{2l_2^2}\nu_2^2 b_q \hat{\theta} P_2^T P_2 + \frac{1}{2}\nu_2^2 + \frac{1}{2} b_5^2 D_q^2. \tag{3.9}$$

By substituting (3.6) and (3.9) into (3.5), it has

$$\begin{aligned}\dot{V}_2 \leq & -k_1\nu_1^2 - \left(k_2 b_5 g_{\vartheta_{\mu q}} - 1\right)\nu_2^2 + \frac{1}{2}\varepsilon_5^2 d^2 + \frac{1}{2} l_1^2 + \frac{1}{2} l_2^2 + \frac{1}{2}\varepsilon_1^2 \\ & + \frac{1}{2}\varepsilon_2^2 + \nu_1(\hat{J} - J)\dot{x}_{1d} + \frac{1}{2} b_5^2 D_q^2 + \frac{1}{2l_1^2}\nu_1^2(\|\phi_1\|^2 - \hat{\theta}) P_1^T P_1 \\ & + \frac{b_q}{2l_2^2}\nu_2^2(\frac{1}{b_q}\|\phi_2\|^2 - \hat{\theta}) P_2^T P_2. \end{aligned} \tag{3.10}$$

Step 3: the tracking error is defined as $z_3 = x_3 - x_{3d}$. From the above equation, we can get $\dot{z}_3 = \dot{x}_3 - \dot{x}_{3d}$. Devise the compensated tracking signal as $\nu_3 = z_3 - \xi_3$. Choose the Lyapunov candidate function as

$$V_3 = V_2 + \frac{1}{2}\nu_3^2.$$

Then the time derivative $\dot{V}_3$ is given by

$$\begin{aligned}\dot{V}_3 \leq & -k_1\nu_1^2 - \left(k_2 b_5 g_{\vartheta_{\mu q}} - 1\right)\nu_2^2 + \frac{1}{2}\varepsilon_5^2 d^2 + \frac{1}{2} l_1^2 + \frac{1}{2} l_2^2 + \frac{1}{2} b_5^2 D_q^2 + \frac{1}{2}\varepsilon_1^2 \\ & + \frac{1}{2}\varepsilon_2^2 + \nu_1(\hat{J} - J)\dot{x}_{1d} + \frac{1}{2l_1^2}\nu_1^2(\|\phi_1\|^2 - \hat{\theta}) P_1^T P_1 + \frac{b_q}{2l_2^2}\nu_2^2(\frac{1}{b_q}\|\phi_2\|^2 \\ & - \hat{\theta}) P_2^T P_2 + \nu_3\left(c_1 x_3 + b_4 x_4 - \dot{x}_{3d} - \dot{\xi}_3\right). \end{aligned} \tag{3.11}$$

Construct the virtual control law α_2 and the compensating signal ξ_3 as

$$\begin{aligned}\alpha_2 &= \frac{1}{b_4}\left(-k_3 z_3 + \dot{x}_{3d} - c_1 x_3\right), \\ \dot{\xi}_3 &= -k_3\xi_3 + b_4\xi_4 + b_4(x_{2,c} - \alpha_2), \end{aligned} \tag{3.12}$$

where $k_3 > 0$, $\xi_3(0) = 0$, $\hat{\theta}$ will be defined later and $x_{2,c}$ is the output of the command filtered with the input signal α_2. Define the error variable and compensating error variable as $z_4 = x_4 - x_{2,c}$, $\nu_4 = z_4 - \xi_4$. Similarly, we have

$$\lim_{t\to\infty}\|\xi_3\| \leq \frac{\mu\rho}{2c_0}.$$

Substituting (3.12) into (3.11), it has

$$\dot{V}_3 = \dot{V}_2 - k_3\nu_3^2 + b_4\nu_3\nu_4. \tag{3.13}$$

Step 4: Similarly, choose the Lyapunov function as

$$V_4 = V_3 + \frac{1}{2}\nu_4^2.$$

Then, $\dot{V}_4$ can be obtained as

$$\dot{V}_4 = \dot{V}_3 - b_4\nu_3\nu_4 + \nu_4(f_4 + b_5 u_d - \dot{\xi}_4),$$

where $f_4(Z) = b_1 x_4 + d_2 x_3 + b_3 x_1 x_2 + b_4\frac{x_2^2}{x_3} + b_4\nu_3 - \dot{x}_{2,c} = \phi_4^T P_4(Z) + \delta_4(Z)$. Similarly,

$$\nu_4 f_4(Z) \le \frac{1}{2l_4^2}\nu_4^2 \|\phi_4\|^2 P_4^T(Z)P_4(Z) + \frac{1}{2}l_4^2 + \frac{1}{2}\nu_4^2 + \frac{1}{2}\varepsilon_4^2.$$

Let

$$u_d = g_{\vartheta_{\mu d}}\vartheta_d + d\left(\vartheta_d\right),$$

with $\dot{\xi}_4 = 0$, and ϑ_d is devised as

$$\vartheta_d = -k_4 z_4 - \frac{1}{2l_4^2}\nu_4\hat{\theta}P_4^T P_4. \tag{3.14}$$

Then, it has

$$b_5 u_d \nu_4 = -k_4 b_5 g_{\vartheta_{\mu d}}\nu_4^2 - \frac{1}{2l_4^2}\nu_4^2 b_5 g_{\vartheta_{\mu d}}\hat{\theta}P_4^T P_4 + \nu_4 b_5 d\left(\vartheta_d\right). \tag{3.15}$$

From $0 < g_{md} < g_{\vartheta_{\mu d}} \le 1$, there exists a positive number b_d that $b_5 g_{\vartheta_{\mu d}} \ge b_d$. By substituting $\nu_4 b_5 d\left(\vartheta_d\right) \le \frac{1}{2}\nu_4^2 + \frac{1}{2}b_5^2 D_d^2$, we have

$$b_5 u_d \nu_4 \le -k_4 b_5 g_{\vartheta_{\mu d}}\nu_4^2 - \frac{1}{2l_4^2}\nu_4^2 b_d\hat{\theta}P_4^T P_4 + \frac{1}{2}\nu_4^2 + \frac{1}{2}b_5^2 D_d^2. \tag{3.16}$$

Define $\theta = \max\{\frac{1}{b}||\phi_1||^2, \frac{1}{b}||\phi_2||^2, \frac{1}{b}||\phi_4||^2\}$, $\tilde{\theta} = \hat{\theta} - \theta, \tilde{J} = \hat{J} - J$, where b =min$\{1, b_q, b_d\}$. Thus, we have

$$\begin{aligned}\dot{V}_4 \le& -k_1\nu_1^2 - \left(k_2 b_5 g_{\vartheta_{\mu q}} - 1\right)\nu_2^2 - k_3\nu_3^2 + \frac{1}{2}l_1^2 + \frac{1}{2}\varepsilon_1^2 + \frac{1}{2}l_4^2 + \frac{1}{2}\varepsilon_4^2 + \frac{1}{2}\varepsilon_5^2 d^2\\ &- \left(k_4 b_5 g_{\vartheta_{\mu d}} - 1\right)\nu_4^2 + \frac{1}{2}b_5^2 D_q^2 + \frac{1}{2}b_5^2 D_d^2 + \frac{1}{2}l_2^2 + \frac{1}{2}\varepsilon_2^2 - \frac{b}{2l_1^2}\nu_1^2\tilde{\theta}P_1^T P_1\\ &- \frac{b}{2l_2^2}\nu_2^2\tilde{\theta}P_2^T P_2 - \frac{b}{2l_4^2}\nu_4^2\tilde{\theta}P_4^T P_4 + \nu_1\tilde{J}\dot{x}_{1d}.\end{aligned} \tag{3.17}$$

Then, consider Lyapunov function of the overall system as

$$V = V_4 + \frac{b}{2r_1}\tilde{\theta}^2 + \frac{1}{2r_2}\tilde{J}^2,$$

where r_1 and r_2 are positive constants. Then, the $\dot{V}$ is given by

$$\begin{aligned}\dot{V} \leq & -k_1\nu_1^2 - \left(k_2 b_5 g_{\vartheta_{\mu q}} - 1\right)\nu_2^2 - k_3\nu_3^2 + \frac{1}{2}l_2^2 + \frac{1}{2}\varepsilon_2^2 - \left(k_4 b_5 g_{\vartheta_{\mu d}} - 1\right)\nu_4^2 + \frac{1}{2}l_4^2 \\ & + \frac{1}{2}\varepsilon_4^2 + \frac{1}{2}\varepsilon_5^2 d^2 + \frac{1}{2}b_5^2 D_q^2 + \frac{1}{2}b_5^2 D_d^2 + \frac{1}{2}l_1^2 + \frac{1}{2}\varepsilon_1^2 + \frac{1}{r_1}\tilde{\theta}(\dot{\hat{\theta}} - \frac{r_1}{2l_1^2}\nu_1^2\tilde{\theta}P_1^T P_1 \\ & \frac{r_1}{2l_1^2}\nu_1^2\tilde{\theta}P_1^T P_1 - \frac{r_1}{2l_2^2}\nu_2^2\tilde{\theta}P_2^T P_2 - \frac{r_1}{2l_4^2}\nu_4^2\tilde{\theta}P_4^T P_4) + \frac{\tilde{J}}{r_2}\left(r_2\nu_1\dot{x}_{1d} + \dot{\hat{J}}\right). \end{aligned} \tag{3.18}$$

Then we select the adaptive laws as

$$\begin{aligned}\dot{\hat{\theta}} &= \frac{r_1}{2l_1^2}\nu_1^2 P_1^T P_1 + \frac{r_1}{2l_2^2}\nu_2^2 P_2^T P_2 + \frac{r_1}{2l_4^2}\nu_4^2 P_4^T P_4 - m_1\hat{\theta}, \\ \dot{\hat{J}} &= -r_2\nu_1\dot{x}_{1d} - m_2\hat{J}, \end{aligned} \tag{3.19}$$

where m_1, m_2, l_1, l_2 and l_4 are positive constants.

Theorem 3.3 *Consider system (3.1)and the given reference signals x_{1d}. If the adaptive law is designed as in (3.19), the adaptive NNs command filtered controllers (3.7), (3.14) can ensure that the resulting tracking errors converge to the origin's small neighborhood. Furthermore, all closed-loop signals of the controlled system are bounded.*

Proof Next, for the stability analysis, we substitute (3.19) into (3.18) holds

$$\begin{aligned}\dot{V} \leq & -k_1\nu_1^2 - \left(k_2 b_5 g_{\vartheta_{\mu q}} - 1\right)\nu_2^2 - \left(k_4 b_5 g_{\vartheta_{\mu d}} - 1\right)\nu_4^2 \\ & -k_3\nu_3^2 + \frac{1}{2}l_1^2 + \frac{1}{2}\varepsilon_1^2 + \frac{1}{2}l_2^2 + \frac{1}{2}\varepsilon_2^2 + \frac{1}{2}l_4^2 + \frac{1}{2}\varepsilon_4^2 \\ & + \frac{1}{2}\varepsilon_5^2 d^2 + \frac{1}{2}b_5^2 D_q^2 + \frac{1}{2}b_5^2 D_d^2 - \frac{m_1\tilde{\theta}\hat{\theta}}{r_1} - \frac{m_2\tilde{J}\hat{J}}{r_2}. \end{aligned} \tag{3.20}$$

Similarly, the following inequalities can be calculated by means of the Young's inequality

$$\begin{aligned} -\tilde{\theta}\hat{\theta} &\leq -\frac{\tilde{\theta}^2}{2} + \frac{\theta^2}{2}, \\ -\tilde{J}\hat{J} &\leq -\frac{\tilde{J}^2}{2} + \frac{J^2}{2}. \end{aligned}$$

Therefore, (3.20) can be rewritten in the following inequality

$$
\begin{aligned}
\dot{V} \leq & -k_1\nu_1^2 - \left(k_2 b_5 g_{\vartheta_{\mu q}} - 1\right)\nu_2^2 - k_3\nu_3^2 - \frac{m_1\tilde{\theta}^2}{2r_1} - \left(k_4 b_5 g_{\vartheta_{\mu d}} - 1\right)\nu_4^2 \\
& -\frac{m_2\tilde{J}^2}{2r_2} + \frac{1}{2}\varepsilon_5^2 d^2 + \frac{1}{2}b_5^2 D_q^2 + \frac{m_2 J^2}{2r_2} + \frac{1}{2}l_1^2 + \frac{1}{2}\varepsilon_1^2 + \frac{1}{2}l_2^2 \\
& +\frac{1}{2}\varepsilon_2^2 + \frac{1}{2}l_4^2 + \frac{m_1\theta^2}{2r_1} + \frac{1}{2}\varepsilon_4^2 \\
\leq & -aV + b,
\end{aligned}
\tag{3.21}
$$

where $a = \min\left\{2k_1/J, 2\left(k_2 b_5 g_{\vartheta_{\mu q}} - 1\right), 2k_3, \ 2\left(k_4 b_5 g_{\vartheta_{\mu d}} - 1\right), m_1, m_2\right\}$ and $b = \frac{1}{2}l_1^2 + \frac{1}{2}\varepsilon_1^2 + \frac{1}{2}l_2^2 + \frac{1}{2}\varepsilon_2^2 + \frac{1}{2}l_4^2 + \frac{1}{2}\varepsilon_4^2 + \frac{1}{2}\varepsilon_5^2 d^2 + \frac{1}{2}b_5^2 D_q^2 + \frac{1}{2}b_5^2 D_d^2 + \frac{m_1\theta^2}{2r_1} + \frac{m_2 J^2}{2r_2}$.

Then, (3.21) implies that

$$
\begin{aligned}
V(t) &\leq (V(t_0) - \frac{b}{a})e^{-a(t-t_0)} + \frac{b}{a} \\
&\leq V(t_0) + \frac{b}{a}, \forall t \geq t_0.
\end{aligned}
\tag{3.22}
$$

All $\nu_i (i = 1, 2, 3, 4)$, $\tilde{J}$ and $\tilde{\theta}$ belong to the compact set Ω.

$$
\Omega = \left\{(\nu_i, \tilde{J}, \tilde{\theta}) | V \leq V(t_0) + \frac{b}{a}, \forall t \geq t_0\right\}.
$$

It means that $\nu_i (i = 1, 2, 3, 4)$, $\tilde{J}$ and $\tilde{\theta}$ are all bounded. Because $z_i = \nu_i + \xi_i$ and $||\xi_i||$ are bounded, the signal z_i will be bounded. By [16], we can obtain $\lim_{t\to\infty} |z_1| \leq \sqrt{\frac{2b}{a} + \frac{\mu\rho}{2c_0}}$. ■

Remark 3.4 According to the definitions of a, b, μ, ρ and c_0, the tracking error z_1 can be very small by selecting sufficiently large r_1, r_2 and small enough $l_1, \ l_2, \ l_4, \varepsilon_1, \varepsilon_2, \varepsilon_4, \varepsilon_5$ after the parameters k_i, m_1, and m_2 are given. ♦

3.4 Simulation Results

To verify the validity of the method proposed in this chapter, the parameters of IM have been chosen as: $J = 0.0586\,\text{kgm}^2$, $R_s = 0.1$, $R_r = 0.15$, $L_s = L_r = 0.0699\text{H}$, $L_m = 0.068\text{H}$, $n_p = 1$. The initial condition is chosen as [0.5, 0.5, 0.5, 0.5]. The reference signals are taken as $x_{1d} = \begin{cases} 50, 0 \leq t < 5, \\ 55, t \geq 5. \end{cases}$ and $x_{3d} = 1$. T_L is chosen as $T_L = 1.0$. The RBF NNs is chosen as follows. The NNs $\phi_1^T P_1(Z)$, $\phi_2^T P_2(Z)$ and

$\phi_4^T P_4(Z)$ contain eleven nodes with centers spaced evenly in the interval $[-9, 9]$ and the partition points are chosen as $9, 7, 5, 3, 1, 0, -1, -3, -5, -7, -9$. Two simulation examples $CFC()$ and y are given to verify the effectiveness of the proposed method. Two simulation examples CFC(I) and CFC(II) are given to verify the effectiveness of the proposed method.

CFC(I): First, the CFC control method is applied to control IM system with the control parameters chosen as: $k_1 = 50, k_2 = 110, k_3 = 150, k_4 = 20, r_1 = r_2 = 0.9, m_1 = m_2 = 0.05, l_1 = l_2 = l_4 = 0.1, \zeta = 0.5, \omega_n = 5000$.

Since u denotes the plant input subject to nonsymmetric saturated nonlinearity described by:

$$u = sat(\vartheta) = \begin{cases} 310, \vartheta \geq 310 \\ \vartheta, -300 < \vartheta < 310 \\ -300, \vartheta \leq -300 \end{cases}.$$

DSC: Then, the DSC control method is utilized to control the IM system, and the control parameters are chosen as same as CFC(I).

CFC(II): Another set of parameters are applied for CFC method. In CFC(II), we select control parameters as: $k_1 = 5, k_2 = 110, k_3 = 150, k_4 = 20, r_1 = r_2 = 0.9, m_1 = m_2 = 0.05, l_1 = l_2 = l_4 = 0.1, \zeta = 0.5, \omega_n = 200$.

Note that Figs. 3.2, 3.4, 3.6, 3.8, 3.10 and 3.12 are the simulation results for CFC(I) method, then the simulation results for DSC technology are shown in Figs. 3.3, 3.5, 3.7, 3.9, 3.11 and 3.13. Among them, Figs. 3.2 and 3.3 show the performance of x_1 and the given reference signal x_{1d}. The tracking error between x_1 and x_{1d} is shown in Figs. 3.4 and 3.5. Similarly, the Figs. 3.6, 3.7, 3.8 and 3.9 is about the tracking performance of x_3 and x_{3d}. Figures 3.10, 3.11, 3.12 and 3.13 shows the input voltage u_q and u_d. The simulation results for CFC(II) can be seen in Figs. 3.14, 3.15, 3.16 and 3.17, in which the curves related to x_1 and x_3 are shown.

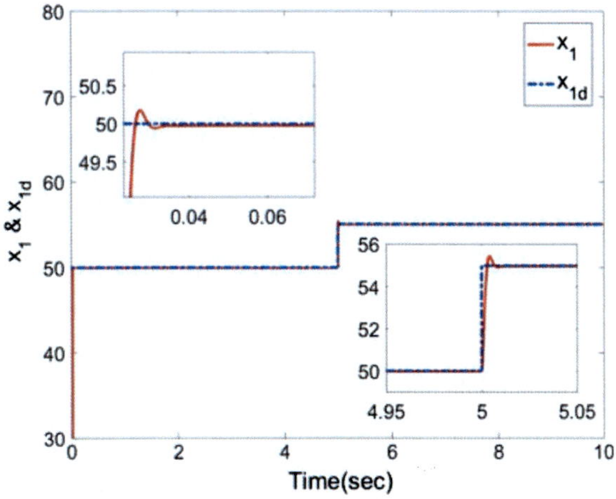

Fig. 3.2 Trajectories of the x_1 and x_{1d} for CFC

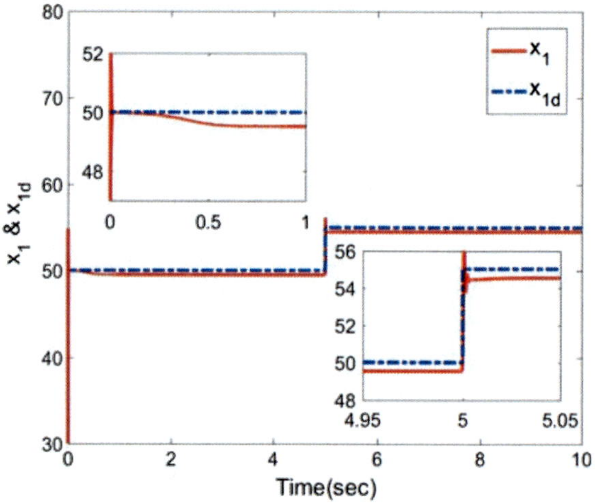

Fig. 3.3 Trajectories of the x_1 and x_{1d} for DSC

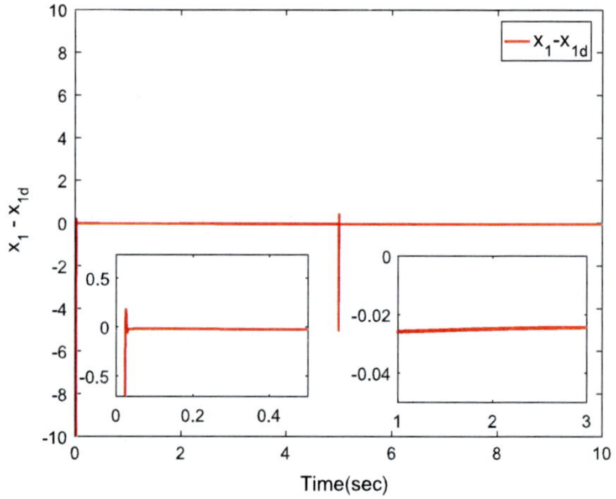

Fig. 3.4 The tracking error of x_1 for CFC(I)

Remark 3.5 From the simulation results in Figs. 3.2, 3.3, 3.4, 3.5, 3.6, 3.7, 3.8, 3.9, 3.10, 3.11, 3.12 and 3.13, we can get that the tracking performances for CFC(I) are better than the DSC method. The preferable tracking performances for CFC(I) are inspired by the errors compensation mechanism in CFC scheme, which is the lack for DSC. ♦

Remark 3.6 In Figs. 3.10, 3.11, 3.12 and 3.13, the actuators designed in this chapter are constrained in a reasonable region, however, the actuators for the DSC method are too large to be used in IM, which means that the method proposed in this chapter is better for practical applications for IM drive systems. ♦

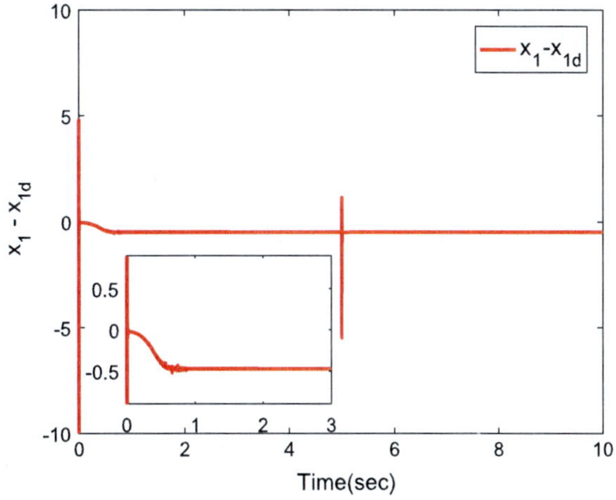

Fig. 3.5 The tracking error of x_1 for DSC

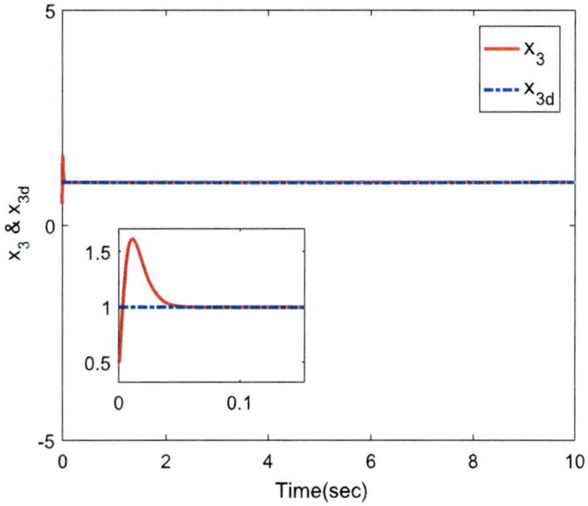

Fig. 3.6 Trajectories of the x_3 and x_{3d} for CFC(I)

Remark 3.7 It can be seen from the simulation results for CFC(I) and CFC(II) that by increasing the ω_n in CFC method, the tracking performance will become better for IM system. However, increasing ω_n will increase the magnitude of the command derivatives system and add the control energy. Therefore, we should choose suitable values of ζ and ω_n according to the filtered results and system control performance. ♦

Fig. 3.7 Trajectories of the x_3 and x_{3d} for DSC

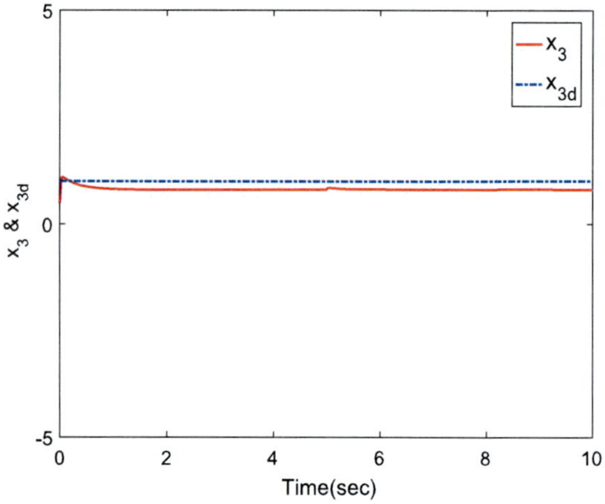

Fig. 3.8 The tracking error of x_3 for CFC(I)

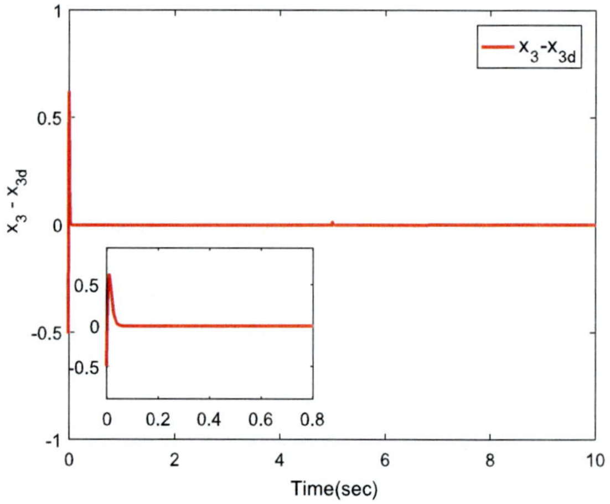

Fig. 3.9 The tracking error of x_3 for DSC

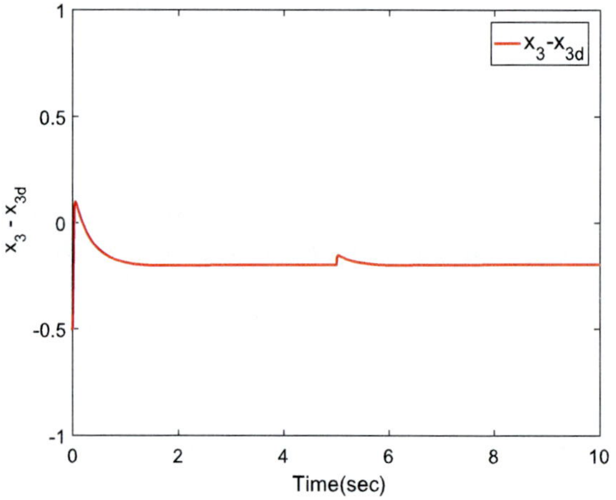

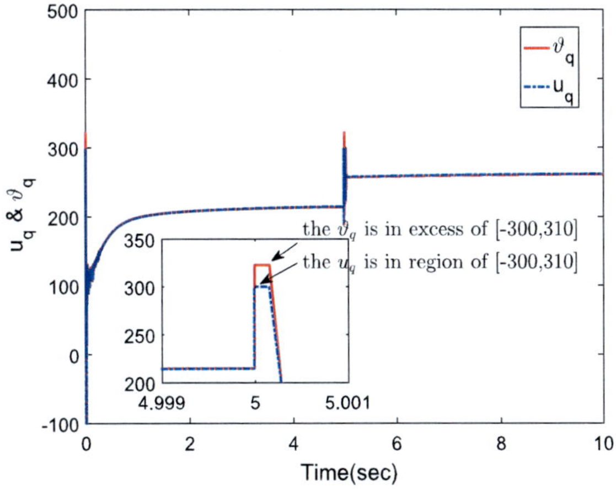

Fig. 3.10 Curve of the control law u_q for CFC(I)

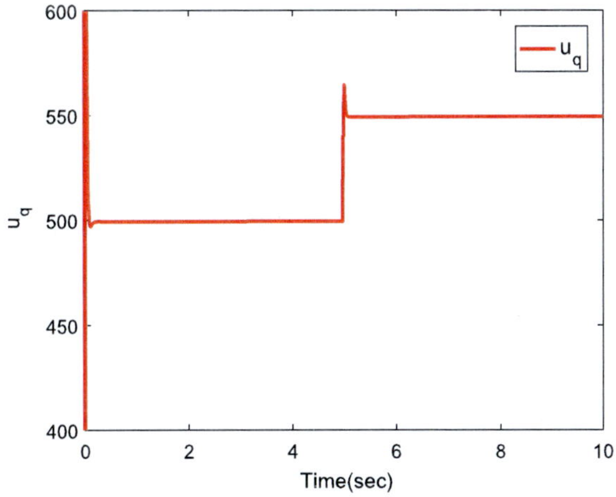

Fig. 3.11 Curve of the control law u_q for DSC

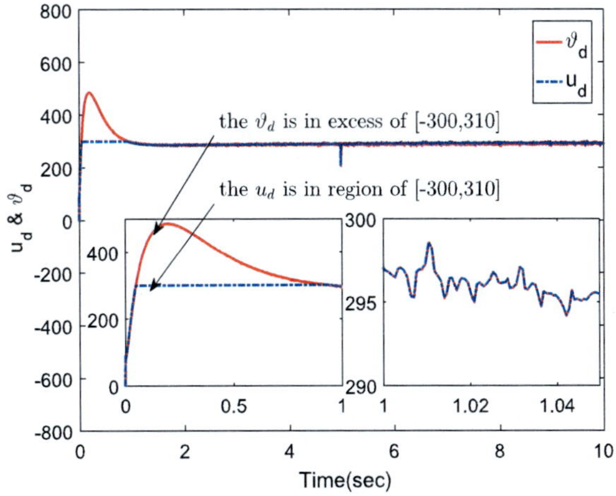

Fig. 3.12 Curve of the control law u_d for CFC(I)

Fig. 3.13 Curve of the control law u_d for DSC

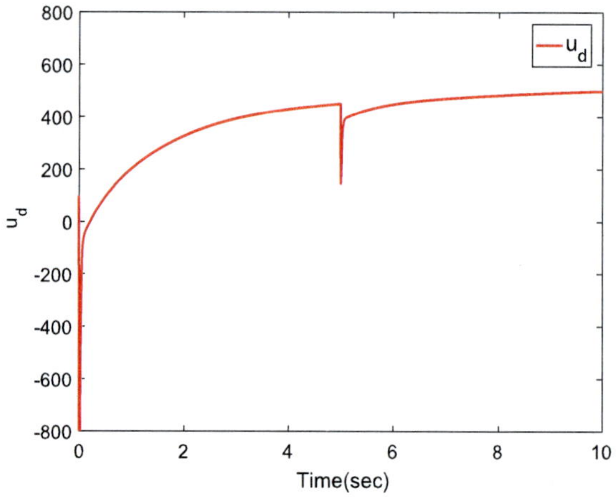

Fig. 3.14 Trajectories of the x_1 and x_{1d} for CFC(I)

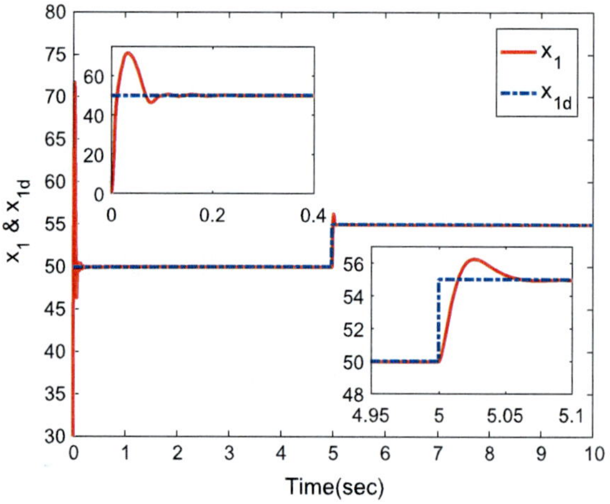

Fig. 3.15 The tracking error of x_1 for CFC(I)

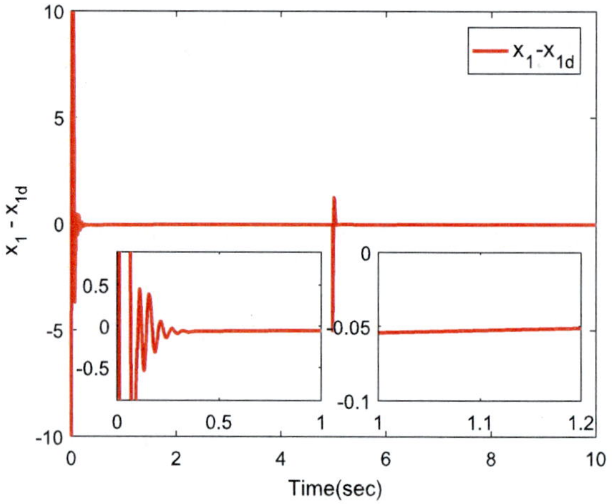

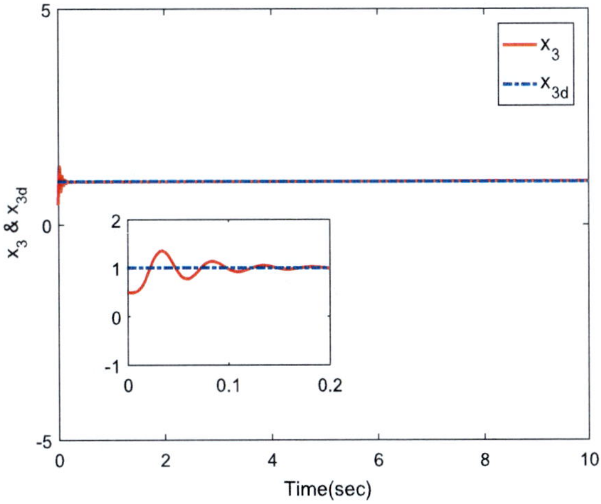

Fig. 3.16 Trajectories of the x_3 and x_{3d} for CFC(II)

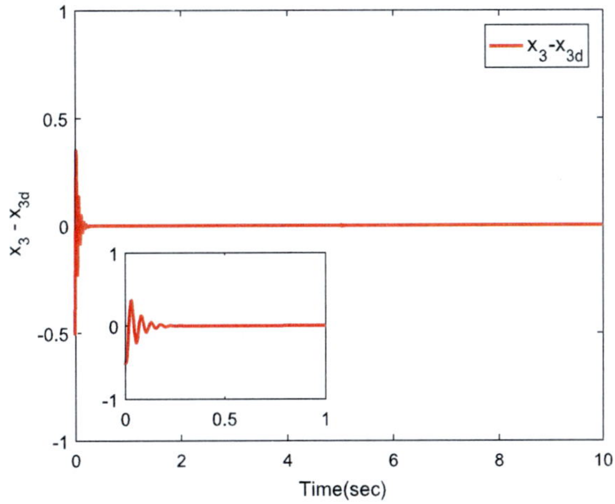

Fig. 3.17 The tracking error of x_3 for CFC(II)

3.5 Conclusion

NNs approximation-based command filtered adaptive control method with input saturation has been devoted for IM. NNs and the adaptive command filtered technology have been respectively utilized to approximate the nonlinearities and solve the "explosion of complexity" problems. This chapter considers the case of input saturation of induction motor, so it is more conducive to practical engineering applications. The simulation results show the superiority of the designed scheme.

References

1. Kojabadi, H.M.: A comparative analysis of different pulse width modulation methods for low cost induction motor drives. Energy Convers. Manage. **52**(1), 136–146 (2011)
2. Reza, C., Islam, M.D., Mekhilef, S.: A review of reliable and energy efficient direct torque controlled induction motor drives. Renew. Sust. Energy Rev. **37**, 919--932 (2014)
3. Nemmour, A.L., Abdessemed, R.: The input-output linearizing control scheme of the doubly-fed induction machine as a wind power generation. Wind Eng. **32**(3), 285–297 (2008)
4. Shen, H., Wu, Z.G., Park, J.H.: Reliable mixed passive and H_∞ filtering for semi-Markov jump systems with randomly occurring uncertainties and sensor failures. Appl. Math. Comput. **25**(17), 3231–3251 (2015)
5. Shen, H., Zhu, Y., Zhang, L., Park, J.H.: Extended dissipative state estimation for Markov jump neural networks with unreliable links. IEEE Trans. Neural Netw. Learn. Syst. **28**(2), 346–358 (2017)
6. Zhao, L., Jia, Y., Yu, J.P., Du, J.: H_∞ sliding mode based scaled consensus control for linear multi-agent systems with disturbances. Appl. Math. Comput. **29**(17), 375–389 (2017)
7. Wang, H.Q., Chen, B., Liu, X.: Robust adaptive fuzzy tracking control for pure-feedback stochastic nonlinear systems with input constraints. IEEE Trans. Cybernet. **43**(6), 2093–2104 (2013)
8. Yang, J., Wu, Z.J.: Stochastic position control for permanent magnet synchronous motor. In: Control and Decision Conference (CCDC), 2010 Chinese. IEEE, pp. 2192–2197 (2010)
9. Cui, G., Xu, S., Zhang, B., Lu, J., Li, Z., Zhang, Z.: Adaptive tracking control for uncertain switched stochastic nonlinear pure-feedback systems with unknown backlash-like hysteresis. J. Frankl. Inst. **354**(4), 1801–1818 (2017)
10. Alonge, F., Filippo, D., Sferlazza, A.: Sensorless control of induction-motor drive based on robust Kalman filter and adaptive speed estimation. IEEE Trans. Ind. Electron. **61**(3), 1444–1453 (2008)
11. Zhou, Z.C., Yu, J.P., Yu, H.S., Lin, C.: Neural-network based discrete-time command filtered adaptive position tracking control for induction motors via backstepping. Neurocomputing **260**, 203–210 (2017)
12. Zhao, L., Jia, Y.: Neural network-based adaptive consensus tracking control for multi-agent systems under actuator fault. Int. J. Syst. Sci. **47**(8), 1931–1942 (2016)
13. Gao, F., Yuan, Y., Wu, Y.: Finite-time stabilization for a class of nonholonomic feedforward systems subject to inputs saturation. ISA Trans. **64**, 193–201 (2016)
14. Hu, Q., Xiao, B., Friswell, M.I.: Robust fault-tolerant control for spacecraft attitude stabilisation subject to input saturation. IET Control Theory Appl. **5**(2), 271–282 (2011)
15. Ge, S.S., Wang, C.: Adaptive NN control of uncertain nonlinear pure-feedback systems. Automatica **38**(4), 671–682 (2002)
16. Dong, W.J., Farrell, J., Polycarpou, M., Djapic, V., Sharma, M.: Command filtered adaptive backstepping. IEEE Trans. Contr. Syst. Technol. **20**(3), 566–580 (2012)

Chapter 4
NNs-Based Discrete-Time Command Filtered Adaptive Control for IM

Considering the problems of parameter uncertainties and load disturbance appeared in induction motor drive systems, a discrete-time command filtered adaptive position tracking control method based on neural networks is proposed in this chapter. First, the euler method is used to describe the discrete-time dynamic mathematical model of induction motor (IM). Next, the neural networks (NNs) technique is employed to approximate the unknown nonlinear functions. Furthermore, the "explosion of complexity" problem and noncausal problem, which emerged in traditional backstepping design, are eliminated by the command filtered control technique. Simulation results prove that tracking error converges to a small neighborhood of the origin and the effectiveness of the proposed approach is illustrated.

4.1 Introduction

In recent years, many control methods have been proposed for IM, such as dynamic surface control [1, 2], Hamiltonian control [3], sliding mode control [4–6], backstepping [7–9], fuzzy logic control [10–13], and some other control methods [14–16]. Unfortunately, all these methods mentioned above were designed for continuous-time IM drive systems. And the design techniques of discrete-time control for IM were seldom mentioned. Considering stability and achievable performances of methods, the discrete-time control systems are generally regarded as superior to continuous-time control systems [14].

In this chapter, the command filtered technique will be applied to nonlinear discrete-time systems with unknown parameters. And the NNs will be used to approximate the uncertain nonlinearities [17–20]. The main merits of the developed scheme can be summed up as follows: (1) The neural networks command filtered backstep-

J. Yu et al., *Intelligent Backstepping Control for the Alternating-Current Drive Systems*, Studies in Systems, Decision and Control 349,
https://doi.org/10.1007/978-3-030-67723-7_4

ping control can solve the problem of "explosion of complexity" to alleviate the online calculational burden; (2) the noncausal problem can be got over by command filtered technique without transforming the system model into a predictor form; (3) the command filtered method can overcome the drawback of traditional method and gain a smaller overshoot. From the above facts, a discrete-time command filtered adaptive control method is developed for position tracking of IM based on neural networks. And the simulation results are provided to illustrate the effectiveness and robustness against the parameter uncertainties and load disturbances.

4.2 Mathematical Model of the IM Drive System

The dynamic mathematical model of IM is described in the well known $(d-q)$ frame as:

$$\begin{cases} \frac{d\Theta}{dt} = \omega \\ \frac{d\omega}{dt} = \frac{n_p L_m}{L_r J}\psi_d i_q - \frac{T_L}{J} \\ \frac{di_q}{dt} = -\frac{L_m^2 R_r + L_r^2 R_s}{\sigma L_s L_r^2} i_q - \frac{L_m n_p}{\sigma L_s L_r}\omega\psi_d - n_p\omega i_d - \frac{L_m R_r}{L_r}\frac{i_q i_d}{\psi_d} + \frac{1}{\sigma L_s}u_q \\ \frac{d\psi_d}{dt} = -\frac{R_r}{L_r}\psi_d + \frac{L_m R_r}{L_r} i_d \\ \frac{di_d}{dt} = -\frac{L_m^2 R_r + L_r^2 R_s}{\sigma L_s L_r^2} i_d + \frac{L_m R_r}{\sigma L_s L_r^2}\psi_d + n_p\omega i_q + \frac{L_m R_r}{L_r}\frac{i_q^2}{\psi_d} + \frac{1}{\sigma L_s}u_d, \end{cases}$$

where $\sigma = 1 - \frac{L_m^2}{L_s L_r}$. n_p, T_L, J, L_m, ω, Θ and ψ_d represent pole pairs, load torque, inertia, the mutual inductance, rotor angular velocity, rotor position and rotor flux linkage. i_d and i_q stand for the $d-q$ axis currents. u_d and u_q are the $d-q$ axis voltages. L_s and R_s mean the inductance, resistance of the stator. R_r and L_r denote the resistance, inductance of the rotor.

By using the Euler method, the IM drivers' dynamic model can be written as:

$$\begin{aligned} x_1(k+1) &= x_1(k) + \Delta_t x_2(k) \\ x_2(k+1) &= x_2(k) + a_1\Delta_t x_3(k)x_4(k) - a_2\Delta_t T_L \\ x_3(k+1) &= (1+b_1\Delta_t)x_3(k) + b_2\Delta_t x_2(k)x_4(k) - \\ &\quad b_3\Delta_t x_2(k)x_5(k) - b_4\Delta_t\frac{x_3(k)x_5(k)}{x_4(k)} + u_q(k)b_5\Delta_t \\ x_4(k+1) &= b_4\Delta_t x_5(k) + x_4(k)(1+c_1\Delta_t) \\ x_5(k+1) &= (1+b_1\Delta_t)x_5(k) + c_2\Delta_t x_4(k) + b_4\Delta_t\frac{x_3^2(k)}{x_4(k)} + \\ &\quad b_3\Delta_t x_2(k)x_3(k) + u_d(k)b_5\Delta_t, \end{aligned} \tag{4.1}$$

where Δ_t is the sampling period and

$$
\begin{aligned}
&x_1(k) = \Theta(k), x_2(k) = \omega(k), x_3(k) = i_q(k),\\
&x_4(k) = \psi_d(k), x_5(k) = i_d(k),\\
&\quad a_1 = \frac{n_p L_m}{L_r J}, a_2 = \frac{1}{J}, b_1 = -\frac{L_m^2 R_r + L_r^2 R_s}{\sigma L_s L_r^2},\\
&\quad b_2 = -\frac{L_m n_p}{\sigma L_s L_r}, b_3 = n_p, b_4 = \frac{L_m R_r}{L_r},\\
&\quad b_5 = \frac{1}{\sigma L_s}, c_1 = -\frac{R_r}{L_r}, c_2 = \frac{L_m R_r}{\sigma L_s L_r^2}.
\end{aligned}
\tag{4.2}
$$

Lemma 4.1 *The command filter is defined as*

$$
\begin{aligned}
z_1(k+1) &= \omega_n z_2(k) \Delta_t + z_1(k),\\
z_2(k+1) &= \{-2\zeta\omega_n z_2(k) - \omega_n (z_1(k) - \alpha_1(k))\} \Delta_t + z_2(k),
\end{aligned}
$$

the input signal $\alpha_1(k)$ satisfies $|\alpha_1(k+1) - \alpha_1(k)| \le \rho_1$, $|\alpha_1(k+2) - 2\alpha_1(k+1) + \alpha_1(k)| \le \rho_2$ for all $k \ge 0$, where ρ_1, ρ_2 are positive constants. And $z_1(0) = \alpha_1(0), z_2(0) = 0$, then for any $\mu > 0$, there exist $\zeta \in (0, 1]$, and $\omega_n > 0$, so we have $|z_1(k) - \alpha_1(k)| \le \mu$ and $\Delta z_1(k) = |z_1(k+1) - z_1(k)|$ is bounded.

The block diagram of the discrete-time neural networks command filtered controller for IM control system is shown as Fig. 4.1. In this paper, the NNs [21] are employed to approximate the continuous function $\varphi(z) : R^q \to R$ as $\hat{\varphi}(z) = \phi^{*T} P(z)$, where $z \in \Omega_z \subset R^q$ is the input variable of the NNs and q is the input dimension, $\phi^* = [\Phi_1^*, \ldots, \Phi_l^*]^T$, is the weight vector with l being the NNs node number. The define of NNs and parameters are shown in [21]. From [21], we know $||P_i(z_i(k))||^2 \le l_i, (i = 1, \cdots, n)$.

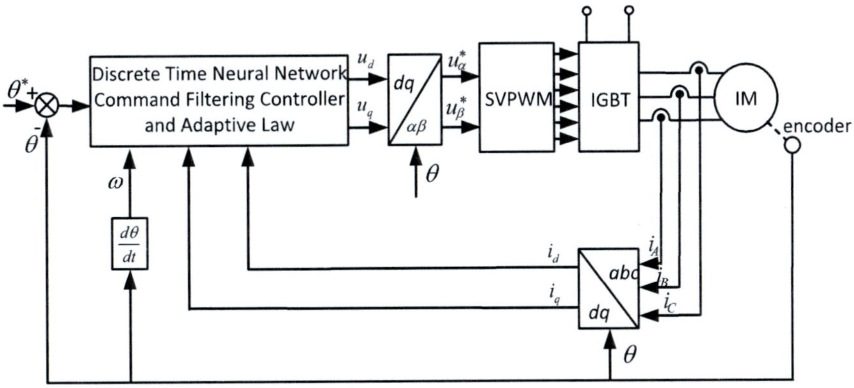

Fig. 4.1 Block diagram of the CFC method for IM

4.3 Discrete-Time Command Filtered Neural Networks Controller Design

In this section, the discrete-time controllers are designed for the IM drive system with backstepping. At each step, one commend filter is needed to filter the virtual control. For $i = 1, 2, 4$, the commend filter is defined as:

$$z_{i,1}(k+1) = \omega_n z_{i,2}(k)\,\Delta_t + z_{i,1}(k), \tag{4.3}$$

$$z_{i,2}(k+1) = \left\{-2\zeta\omega_n z_{i,2}(k) - \omega_n\left(z_{i,1}(k) - \alpha_i(k)\right)\right\}\Delta_t + z_{i,2}(k), \tag{4.4}$$

where $\alpha_i(k)$ is the input and $z_{i,1}(k)$ is the output of the filter. The initial condition of the filter is $z_{i,1}(0) = \alpha_i(0)$, and $z_{i,2}(0) = 0$.

Step 1: The tracking error variable is defined as $e_1(k) = x_1(k) - x_{1d}(k)$ with the desired signal $x_{1d}(k)$. According to the Eq. (4.1), we can obtain $e_1(k+1) = \Delta_t x_2(k) + x_1(k) - x_{1d}(k+1)$. Define the Lyapunov function as $V_1(k) = \frac{1}{2}e_1^2(k)$, and the difference of $V_1(k)$ can be written as

$$\Delta V_1(k) = -\frac{1}{2}e_1^2(k) + \frac{1}{2}[x_1(k) + \Delta_t x_2(k) - x_{1d}(k+1)]^2.$$

The virtual control law $\alpha_1(k)$ is chosen as

$$\alpha_1(k) = \frac{-x_1(k) + x_{1d}(k+1)}{\Delta_t}. \tag{4.5}$$

Define $e_2(k) = x_2(k) - x_{1c}(k)$, where $x_{ic}(k) = z_{i,1}(k)$, $(i = 1, 2, 4)$ as the outputs of command filters. By using (4.5), $\Delta V_1(k)$ can be given as

$$\Delta V_1(k) = \frac{1}{2}\left[e_2(k) + x_{1c}(k) - \alpha_1(k)\right]^2 \Delta_t^2 - \frac{1}{2}e_1^2(k).$$

Step 2: By use of the Eq. (4.4), $e_2(k+1)$ is obtained as $e_2(k+1) = a_1\Delta_t x_3(k) x_4(k) + x_2(k) - a_2\Delta_t T_L - x_{1c}(k+1)$. Define the Lyapunov function as $V_2(k) = V_1(k) + \frac{1}{2}e_2^2(k)$. Furthermore, differencing $V_2(k)$ yields

$$\begin{aligned}\Delta V_2(k) = {} & -\frac{1}{2}e_2^2(k) + \frac{1}{2}\left[x_2(k) + a_1\Delta_t x_3(k)x_4(k)\right.\\ & \left.-a_2\Delta_t T_L - x_{1c}(k+1)\right]^2 + \Delta V_1(k).\end{aligned}$$

In this chapter, due to the parameter T_L being bounded in the practice system, we assume the T_L is unknown, but its upper bound is $d > 0$. Namely, $0 \le T_L \le d$. The virtual control law $\alpha_2(k)$ is constructed as

$$\alpha_2(k) = \frac{-x_2(k) + x_{1c}(k+1) + a_2 \Delta_t d}{a_1 \Delta_t x_4(k)}, \tag{4.6}$$

where $x_{1c}(k+1)$ can be calculated by (4.3).

Remark 4.1 It can be seen that the virtual controller $\alpha_2(k)$ contains variable $x_{1c}(k+1)$, which covers future information. And the controller will contain more future information when we continue to design the real controller via backstepping, which is impossible in practice and this drawback was named the noncausal problem [10]. The existing result to overcome this problem is transforming the systems into a predictor form, which will add the control complexity. In this chapter, $x_{1c}(k+1)$ can be calculated by the variable value of the previous step from the command filter, then the noncausal problem can be solved.

Define $e_3(k) = x_3(k) - x_{2c}(k)$. By use of (4.6), we can obtain

$$\Delta V_2(k) \leq \frac{1}{2}\left[e_3(k) + x_{2c}(k) - \alpha_2(k)\right]^2 a_1^2 \Delta_t^2 x_4^2(k) - \frac{1}{2} e_2^2(k) + \Delta V_1(k).$$

Step 3: According to (4.1), we can obtain $e_3(k+1) = b_5 \Delta_t u_q(k) + f_3(k)$, where $f_3(k) = (1 + b_1 \Delta_t) x_3(k) + b_2 \Delta_t x_2(k) x_4(k) - b_3 \Delta_t x_2(k) x_5(k) - b_4 \Delta_t \frac{x_3(k) x_5(k)}{x_4(k)} - x_{2c}(k+1)$. Define the Lyapunov function as $V_3(k) = \frac{1}{2} e_3^2(k) + V_2(k)$. Obviously, the difference of $V_3(k)$ is computed by

$$\Delta V_3(k) = \frac{1}{2}\left[b_5 \Delta_t u_q(k) + f_3(k)\right]^2 - \frac{1}{2} e_3^2(k) + \Delta V_2(k).$$

Remark 4.2 It can be obtained that the nonlinear terms $b_2 \Delta_t x_2(k) x_4(k)$, $b_3 \Delta_t x_2(k) x_5(k)$ and $b_4 \Delta_t \frac{x_3(k) x_5(k)}{x_4(k)}$ are in $f_3(k)$, which will add the complexity and difficulty during the design procedure of real controller $u_q(k)$ and backstepping. Herein, NNs are used to approximate the nonlinear function $f_3(k)$ to simplify the control signal structure. As shown later, the design procedure of real controller $u_q(k)$ becomes simpler and the structure of $u_q(k)$ is briefer and more practical.

By using the NNs, for any $\varepsilon_3 > 0$, there exists a NNs $\phi_3^T P_3(z_3(k))$ such that

$$f_3(k) = \phi_3^T P_3(z_3(k)) + \epsilon_3, \tag{4.7}$$

where $z_3(k) = [x_2(k), x_3(k), x_4(k), x_5(k), x_{2c}(k+1)]^T$. ϵ_3 is the approximation error, and $|\epsilon_3| \leq \varepsilon_3$. At this present stage, the adaptive law $\hat{\eta}_3(k+1)$ and control law $u_q(k)$ are defined as

$$u_q(k) = -\frac{1}{b_5 \Delta_t} \hat{\eta}_3(k) ||P_3(z_3(k))||, \tag{4.8}$$

$$\hat{\eta}_3(k+1) = \hat{\eta}_3(k) + \gamma_3 ||P_3(z_3(k))|| e_3(k+1) - \delta_3 \hat{\eta}_3(k), \tag{4.9}$$

where γ_3 and δ_3 are positive parameters. In general, ϕ_3 is bounded and unknown. Let $||\phi_3|| = \eta_3$, where $\eta_3 > 0$ is an unknown constant. Use $\hat{\eta}_3(k)$ as the estimation of η_3 and define $\tilde{\eta}_3(k) = \eta_3 - \hat{\eta}_3(k)$, where $\tilde{\eta}_3(k)$ is the estimate error.

By using equality (4.8), we have

$$\Delta V_3(k) \leq ||P_3(z_3(k))||^2 \tilde{\eta}_3^2(k) + \varepsilon_3^2 - \frac{1}{2} e_3^2(k) + \Delta V_2(k). \tag{4.10}$$

Step 4: For the desired signal $x_{4d}(k)$, the tracking error is defined as $e_4(k) = x_4(k) - x_{4d}(k)$. According to the Eq. (4.1), we can obtain $e_4(k+1) = (1 + c_1 \Delta_t) x_4(k) + b_4 \Delta_t x_5(k) - x_{4d}(k+1)$. Consider the Lyapunov function as $V_4(k) = \frac{1}{2} e_4^2(k) + V_3(k)$. Furthermore, differencing $V_4(k)$ yields

$$\Delta V_4(k) = \frac{1}{2} [(1 + c_1 \Delta_t) x_4(k) - x_{4d}(k+1) + b_4 \Delta_t x_5(k)]^2 \\ -\frac{1}{2} e_4^2(k) + \Delta V_3(k). \tag{4.11}$$

The virtual control law $\alpha_4(k)$ is defined as

$$\alpha_4(k) = \frac{1}{b_4 \Delta_t} [-(1 + c_1 \Delta_t) x_4(k) + x_{4d}(k+1)]. \tag{4.12}$$

Using (4.12), $\Delta V_4(k)$ can be rewritten as

$$\Delta V_4(k) \leq \frac{1}{2} b_4^2 \Delta_t^2 [e_5(k) + x_{4c}(k) - \alpha_4(k)]^2 - \frac{1}{2} e_4^2(k) + \Delta V_3(k), \tag{4.13}$$

with $e_5(k) = x_5(k) - x_{4c}(k)$.

Step 5: By use of the Eq. (4.1), we have $e_5(k+1) = f_5(k) + b_5 \Delta_t u_d(k)$, where $f_5(k) = (1 + b_1 \Delta_t) x_5(k) + c_2 \Delta_t x_4(k) + b_3 \Delta_t x_2(k) x_3(k) + b_4 \Delta_t \frac{x_3^2(k)}{x_4(k)} - x_{4c}(k+1)$. Define the Lyapunov function as $V_5(k) = \frac{P}{2} e_5^2(k) + V_4(k)$, where $P > 0$. Consequently, we can obtain the difference of $V_5(k)$ as

$$\Delta V_5(k) = \frac{P}{2} [f_5(k) + b_5 \Delta_t u_d(k)]^2 - \frac{P}{2} e_5^2(k) + \Delta V_4(k). \tag{4.14}$$

Similarly, there exists a NNs $\phi_5^T P_5(z_5(k))$ such that $f_5(k)$ can be approximated as

$$f_5(k) = \phi_5^T P_5(z_5(k)) + \epsilon_5,$$

where $z_5(k) = [x_2(k), x_3(k), x_4(k), x_5(k), x_{4c}(k+1)]^T$. ϵ_5 is the approximation error, and $|\epsilon_5| \le \varepsilon_5$. Now we define the adaptive law $\hat{\eta}_5(k+1)$ and control law $u_d(k)$ as following equations

$$u_d(k) = -\frac{1}{b_5 \Delta_t}\hat{\eta}_5(k)||P_5(z_5(k))||, \tag{4.15}$$

$$\hat{\eta}_5(k+1) = \hat{\eta}_5(k) + \gamma_5||P_5(z_5(k))||e_5(k+1) - \delta_5\hat{\eta}_5(k), \tag{4.16}$$

where γ_5 and δ_5 are positive parameters. In general, ϕ_5 is bounded and unknown. Let $||\phi_5|| = \eta_5$, where $\eta_5 > 0$ is an unknown constant. Using $\hat{\eta}_5(k)$ as the estimation of η_5 and we have $\tilde{\eta}_5(k) = \eta_5 - \hat{\eta}_5(k)$. By using $|x_{i_c}(k) - \alpha_i(k)| \le \mu_i$, $(i = 1, 2, 4)$ and substituting (4.15) into (4.14) results in

$$\begin{aligned}
\Delta V_5(k) &\le P\tilde{\eta}_5^2(k)||P_5(z_5(k))||^2 - \frac{1}{2}Pe_5^2(k) + P\varepsilon_5^2 + \Delta V_4(k) \\
&\le P\tilde{\eta}_5^2(k)||P_5(z_5(k))||^2 - \frac{1}{2}Pe_5^2(k) - \frac{1}{2}e_4^2(k) \\
&\quad + b_4^2\Delta_t^2 e_5^2(k) + b_4^2\Delta_t^2\mu_4^2 + P\varepsilon_5^2 \\
&\quad + \tilde{\eta}_3^2(k)||P_3(z_3(k))||^2 - \frac{1}{2}e_3^2(k) + \varepsilon_3^2 \\
&\quad + a_1^2\Delta_t^2 e_3^2(k)x_4^2(k) + a_1^2\Delta_t^2 x_4^2(k)\mu_2^2 \\
&\quad + \Delta_t^2 e_2^2(k) + \Delta_t^2\mu_1^2 - \frac{1}{2}e_1^2(k) - \frac{1}{2}e_2^2(k).
\end{aligned} \tag{4.17}$$

Theorem 4.3 *Consider system (4.1) and reference signals x_{1d} and x_{4d} under the designed controller. If the virtual control signals are constructed as in (4.5), (4.6) and (4.12), the adaptive laws are designed as in (4.9) and (4.16), then we choose the adaptive neural networks controllers (4.8) and (4.15) such that the resulting tracking errors converge to the origin's small neighborhood. Also, all closed-loop signals of the controlled system are bounded.*

Proof In order to prove that all the signals are bounded in the system, the Lyapunov function is defined as $V(k) = V_5(k) + \frac{1}{2\gamma_3}\tilde{\eta}_3^2(k) + \frac{P}{2\gamma_5}\tilde{\eta}_5^2(k)$, where γ_3 and γ_5 are positive constants. Then, $\Delta V(k)$ can be rewritten as

$$\Delta V(k) = \Delta V_5(k) + \frac{1}{2\gamma_3}\left[\tilde{\eta}_3^2(k+1) - \tilde{\eta}_3^2(k)\right] + \frac{P}{2\gamma_5}\left[\tilde{\eta}_5^2(k+1) - \tilde{\eta}_5^2(k)\right]. \tag{4.18}$$

Based on $||P_i(z_i(k))||^2 \le l_i$, $(i = 3, 5)$ and Eqs. (4.7) and (4.8), where l_i denotes the neurons used, we can get

$$e_3^2(k+1) \leq 2\tilde{\eta}_3^2(k)\, l_3 + 2\varepsilon_3^2.$$

Similarly, we have

$$e_5^2(k+1) \leq 2\tilde{\eta}_5^2(k)\, l_5 + 2\varepsilon_5^2.$$

By using $\tilde{\eta}_i(k) = \eta_i - \hat{\eta}_i(k)$, we can obtain

$$\tilde{\eta}_i^2(k+1) - \tilde{\eta}_i^2(k) = \eta_i^2 + \hat{\eta}_i^2(k+1) - 2\eta_i\hat{\eta}_i(k+1) - \tilde{\eta}_i^2(k), \tag{4.19}$$

$$\begin{aligned}\hat{\eta}_i^2(k+1) = {} & \gamma_i^2 e_i^2(k+1)||P_i(z_i(k))||^2 + (1-\delta_i)^2\hat{\eta}_i^2(k) + \\ & 2(1-\delta_i)\gamma_i||P_i(z_i(k))||e_i(k+1)\hat{\eta}_i(k).\end{aligned} \tag{4.20}$$

Substituting (4.20) into (4.19) yields

$$\begin{aligned}\tilde{\eta}_i^2(k+1) - \tilde{\eta}_i^2(k) = {} & \eta_i^2 + \gamma_i^2 e_i^2(k+1)||P_i(z_i(k))||^2 + \\ & (1-\delta_i)^2\hat{\eta}_i^2(k) - 2(1-\delta_i)\eta_i\hat{\eta}_i(k) + \\ & 2(1-\delta_i)\gamma_i||P_i(z_i(k))||e_i(k+1)\hat{\eta}_i(k) \\ & - \tilde{\eta}_i^2(k) - 2\gamma_i||P_i(z_i(k))||e_i(k+1)\eta_i.\end{aligned} \tag{4.21}$$

According to Young's inequality [10], we can obtain

$$\begin{aligned}2\gamma_i \|S_i(z_i(k))\|\, e_i(k+1)\,\hat{\eta}_i(k) &\leq \gamma_i^2 e_i^2(k+1)\, l_i + \hat{\eta}_i^2(k) \\ -2||S_i(z_i(k))||e_i(k+1)\eta_i &\leq e_i^2(k+1)l_i + \eta_i^2 \\ \gamma_i^2 e_i^2(k+1)||S_i(z_i(k))||^2 &\leq \gamma_i^2 e_i^2(k+1)l_i \\ -2\eta_i\hat{\eta}_i(k) &\leq \eta_i^2 + \hat{\eta}_i^2(k).\end{aligned} \tag{4.22}$$

Then, substituting (4.22) into (4.21), we have

$$\begin{aligned}\tilde{\eta}_i^2(k+1) - \tilde{\eta}_i^2(k) \leq {} & (\gamma_i - \delta_i + 2)\eta_i^2 + (\delta_i^2 - 4\delta_i + 3)\hat{\eta}_i^2(k) \\ & + (4\gamma_i^2 l_i^2 - 2\gamma_i^2\delta_i l_i^2 + 2\gamma_i l_i^2 - 1)\tilde{\eta}_i^2(k) \\ & + (4\gamma_i^2 l_i - 2\gamma_i^2\delta_i l_i + 2\gamma_i l_i)\varepsilon_i^2,\ (i = 3, 5)\end{aligned} \tag{4.23}$$

Define $x_4^2(k) \leq N$, where $N > 0$ is a constant. Substituting (4.23) and (4.17) into (4.18), one has

$$\begin{aligned}\Delta V \leq {} & -e_5^2(k)\left(\frac{P}{2} - b_4^2\Delta_t^2\right) - e_3^2(k)\left(\frac{1}{2} - a_1^2\Delta_t^2 N\right) - e_2^2(k)\left(\frac{1}{2} - \Delta_t^2\right) \\ & - \frac{1}{2}e_4^2(k) - \frac{1}{2}e_1^2(k) + \frac{1}{2\gamma_3}\left[\hat{\eta}_3^2(k)\left(\delta_3^2 - 4\delta_3 + 3\right) + \beta_3\right. \\ & \left. + \tilde{\eta}_3^2(k)\left(4\gamma_3^2 l_3^2 - 2\gamma_3^2\delta_3 l_3^2 + 2\gamma_3 l_3^2 + 2\gamma_3 l_3 - 1\right)\right]\end{aligned}$$

$$+\frac{P}{2\gamma_5}\left[\hat{\eta}_5^2(k)\left(\delta_5^2-4\delta_5+3\right)+\beta_5\right.$$
$$\left.+\tilde{\eta}_5^2(k)\left(4\gamma_5^2l_5^2-2\gamma_5^2\delta_5l_5^2+2\gamma_5l_5^2+2\gamma_5l_5-1\right)\right],$$

$$\begin{aligned}\beta_3 &= \left(4\gamma_3^2l_3-2\gamma_3^2\delta_3l_3+2\gamma_3l_3+2\gamma_3\right)\varepsilon_3^2+2\gamma_3a_1^2\Delta_t^2N\mu_2^2\\&+(\gamma_3-\delta_3+2)\,\eta_3^2+\gamma_3\Delta_t^2\mu_1^2,\\ \beta_5 &= \left(4\gamma_5^2l_5-2\gamma_5^2\delta_5l_5+2\gamma_5l_5+2\gamma_5\right)\varepsilon_5^2+\frac{2\gamma_5}{P}b_4^2\Delta_t^2\mu_4^2\\&+(\gamma_5-\delta_5+2)\,\eta_5^2+\gamma_5\Delta_t^2\mu_1^2.\end{aligned}$$

By selecting an appropriate sampling period Δ_t and parameter P, we can guarantee $\frac{P}{2}-b_4^2\Delta_t^2>0$, $\frac{1}{2}-a_1^2\Delta_t^2N>0$ and $\frac{1}{2}-\Delta_t^2>0$. We ensure these reference parameters as following inequalities: $4\gamma_i^2l_i^2-2\gamma_i^2\delta_il_i^2+2\gamma_il_i^2+2\gamma_il_i-1<0$ and $\delta_i^2-4\delta_i+3<0$, $(i=3,5)$. Once the errors $|e_5(k)|>\sqrt{\frac{P\beta_5}{-2\gamma_5b_4^2\Delta_t^2+P\gamma_5}}$ and $|e_3(k)|>\sqrt{\frac{\beta_3}{\gamma_3-2\gamma_3a_1^2\Delta_t^2N}}$, then we can obtain $\Delta V(k)\le 0$, $\lim_{k\to\infty}||x_1(k)-x_{1d}(k)||\le\xi$ for $\xi>0$ is a small constant. ■

Remark 4.4 From the definitions of ξ, the tracking error can be very small by choosing small enough $\varepsilon_3, \varepsilon_5$ and sufficiently large γ_3, γ_5 after the parameters δ_3 and δ_5 are defined.

4.4 Simulation Results

To prove the effectiveness of the control method proposed in this chapter, a simulation is run and the parameters of the IM are chosen as:

$$J=0.0586\,\text{Kgm}^2,\ R_s=0.1\Omega,\ R_r=0.15\Omega,$$
$$L_m=0.068\text{H},\ n_p=1,\ L_s=L_r=0.0699\text{H},$$

and $x_1(0)=x_2(0)=x_3(0)=x_5(0)=0$, $x_4(0)=1.8$ are defined as the initial condition for the IM in the simulation. The reference signals are selected as $x_{1d}(k)=2\cos(\Delta_tk\pi/2)$, $x_{4d}(k)=1$, and take T_L as:

$$T_L=\begin{cases}0.5,\ 0\le k<2000,\\1.0,\ k\ge 2000.\end{cases}$$

Simulation for the command filtered controller in this chapter. Considering the system efficiency and control performance, $\Delta_t=0.0025$s is selected as the sampling period. The design parameters are chosen as $\delta_3 = 0.87$, $\delta_5 = 0.0021$, $\zeta = 0.9$, $\omega_n = 200$, $\gamma_3 = 0.98$ and $\gamma_5 = 0.25$.

Simulation for the traditional controller without command filter. The design parameters are chosen as $\delta_3 = 0.77$, $\delta_5 = 0.0031$, $\gamma_3 = 0.98$, $\gamma_5 = 0.36$, and other parameters are the same as CFC.

Choose the NNs membership functions for the NNs $\phi_3^T P_3(z_3(k))$ as

$$\Phi_1^* = \exp\left[\frac{-(x(k)+3)^2}{40}\right], \Phi_2^* = \exp\left[\frac{-(x(k)+2)^2}{40}\right],$$
$$\Phi_3^* = \exp\left[\frac{-x^2(k)}{40}\right], \Phi_4^* = \exp\left[\frac{-(x(k)-2)^2}{40}\right],$$
$$\Phi_5^* = \exp\left[\frac{-(x(k)-3)^2}{40}\right].$$

Choose the NNs membership functions for the NNs $\phi_5^T P_5(z_5(k))$ as

$$\Phi_1^* = \exp\left[\frac{-(x(k)+10)^2}{40}\right], \Phi_2^* = \exp\left[\frac{-(x(k)+5)^2}{40}\right],$$
$$\Phi_3^* = \exp\left[\frac{-x^2(k)}{40}\right], \Phi_4^* = \exp\left[\frac{-(x(k)-5)^2}{40}\right],$$
$$\Phi_5^* = \exp\left[\frac{-(x(k)-10)^2}{40}\right].$$

By using the proposed control method, it can be seen that the results of simulation in Figs. 4.2, 4.3, 4.4, 4.5, 4.6, 4.7, 4.8, 4.9, 4.10, 4.11, 4.12, 4.13, 4.14 and 4.15, where Figs. 4.2, 4.4, 4.6, 4.8, 4.10, 4.12 and 4.14 display the command filtered method and Fig. 4.3, 4.5, 4.7, 4.9, 4.11, 4.13 and 4.15 reflect the traditional control method. Figures 4.2, 4.3 and Figs. 4.14, 4.15 display the trajectories of $x_1(k)$, $x_{1d}(k)$ and $x_4(k)$, $x_{4d}(k)$. In Figs. 4.2, 4.3 and Figs. 4.14, 4.15 the dashed line represents $x_{1d}(k)$ and $x_{4d}(k)$, the solid line represents $x_1(k)$ and $x_4(k)$. These figures show that the desired reference signals can be tracked well by the system output. From Figs. 4.4 and 4.5, we can know the tracking errors of the simulation converge to a small neighborhood of the area. Obviously, the overshoot in Fig. 4.5 is much larger than that in Fig. 4.4. Figures 4.6, 4.7 and Figs. 4.8, 4.9 represent the simulation results of $u_q(k)$ and $u_d(k)$. And $i_d(k)$, $i_q(k)$ are shown in Figs. 4.10, 4.11 and Figs. 4.12, 4.13. From Figs. 4.6, 4.7, 4.8, 4.9, 4.10, 4.11, 4.12 and 4.13, we know that $u_q(k)$, $u_d(k)$ and $i_d(k)$, $i_q(k)$ are bounded into a certain area. From the simulation, we know that the controllers have better robustness to resistance load disturbances and parameter changes.

Remark 4.5 From the simulations, it can be clearly seen that both two kinds of methods can gain good control effects. Compared the above two sets of simulation results, we can see that the approach proposed in this chapter can achieve better tracking effect and the performance of the proposed control method to reject overshoots is better than the traditional backstepping approach.

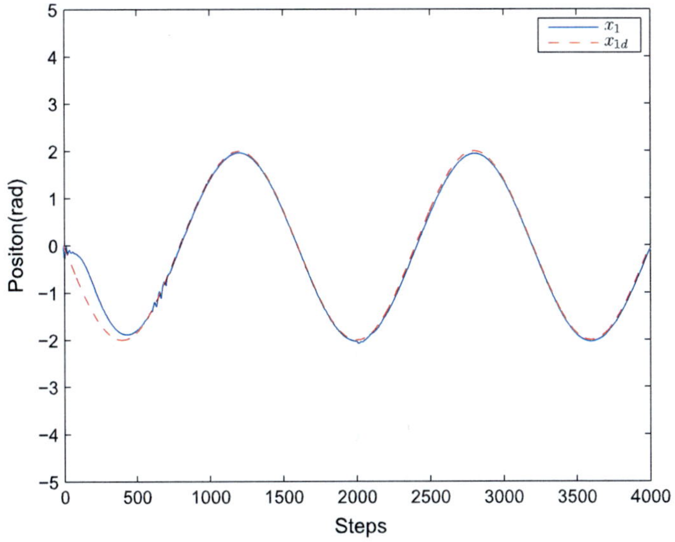

Fig. 4.2 Trajectories of the x_1 and x_{1d} for CFC

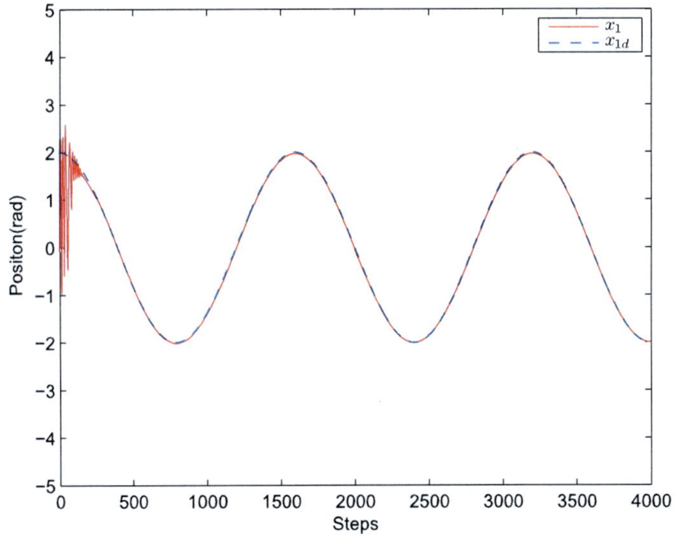

Fig. 4.3 Trajectories of the x_1 and x_{1d} without CFC

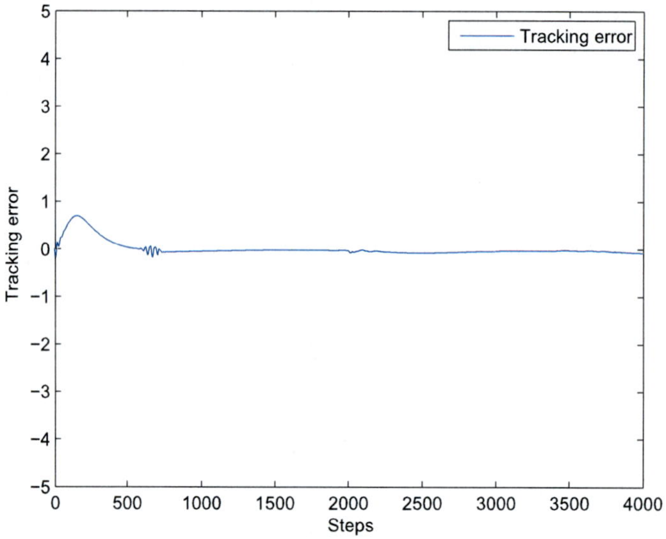

Fig. 4.4 The tracking error of x_1 and x_{1d} for CFC

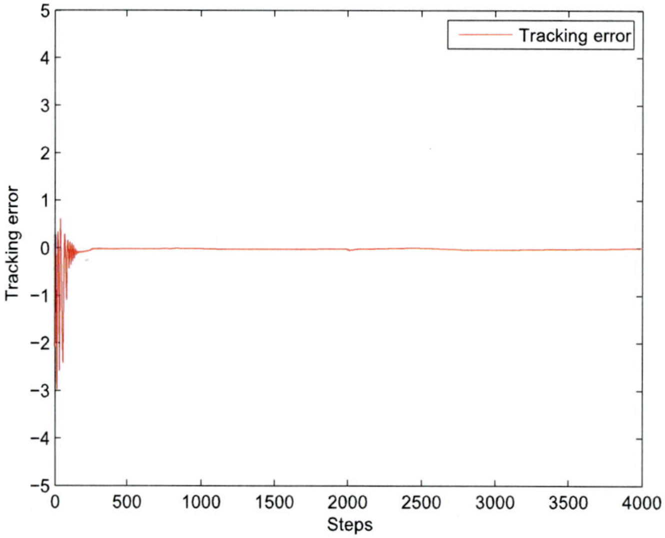

Fig. 4.5 The tracking error of x_1 and x_{1d} without CFC

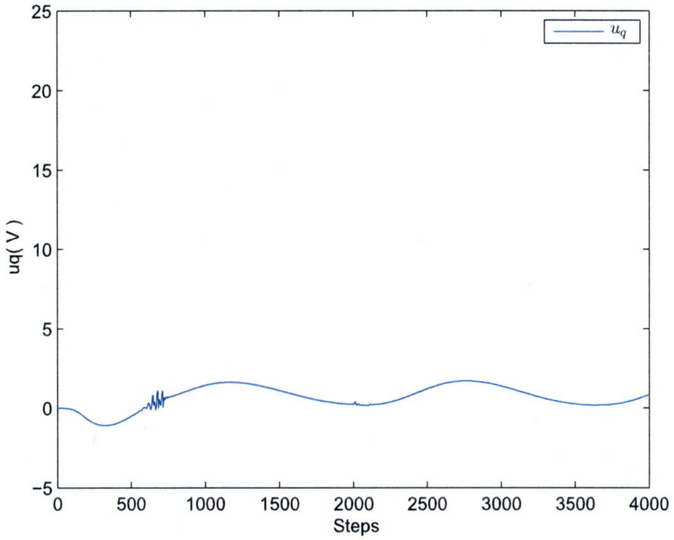

Fig. 4.6 Curve of the u_q for CFC

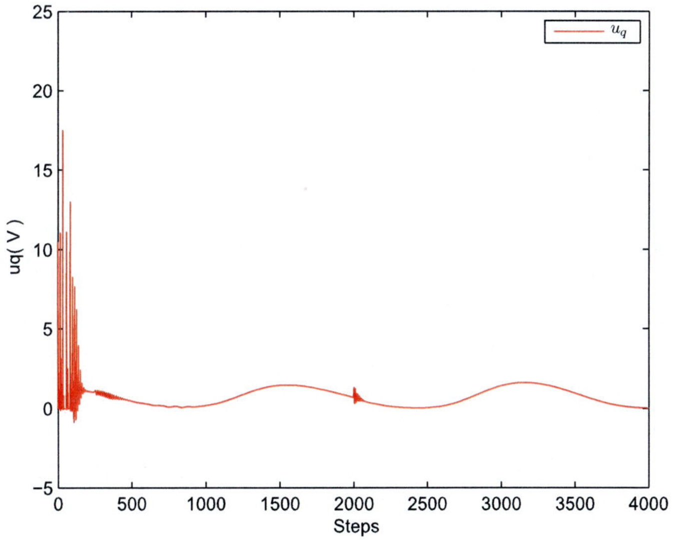

Fig. 4.7 Curve of the u_q without CFC

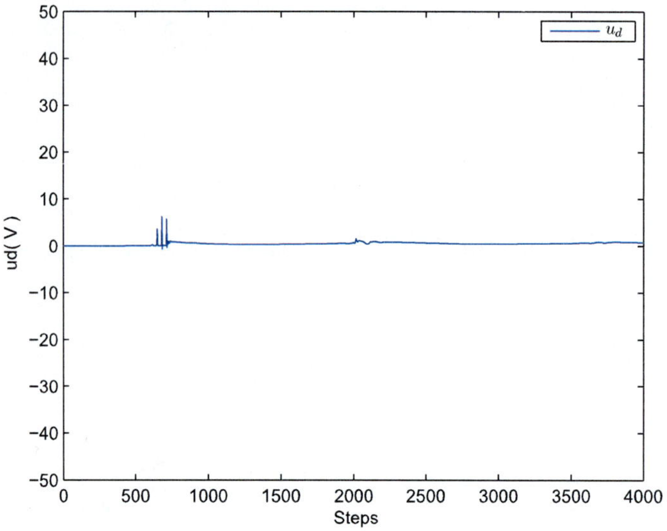

Fig. 4.8 Curve of the u_d for CFC

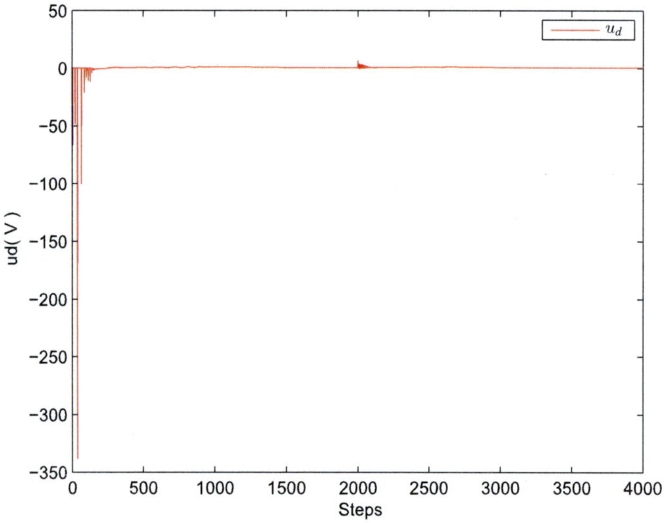

Fig. 4.9 Curve of the u_d without CFC

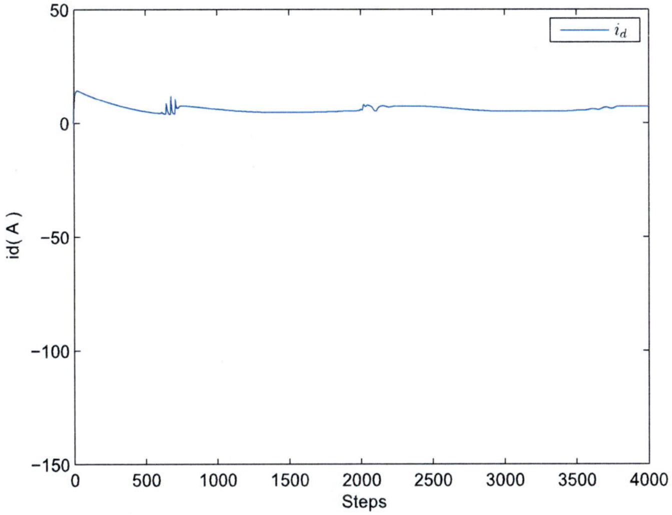

Fig. 4.10 Curve of the i_d for CFC

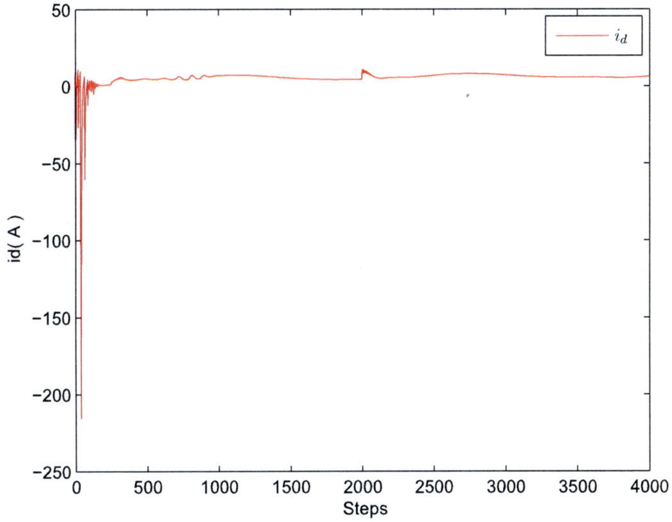

Fig. 4.11 Curve of the i_d without CFC

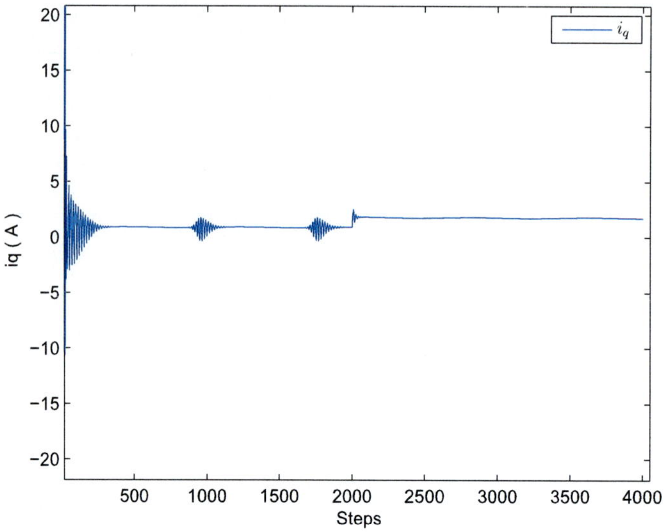

Fig. 4.12 Curve of the i_q for CFC

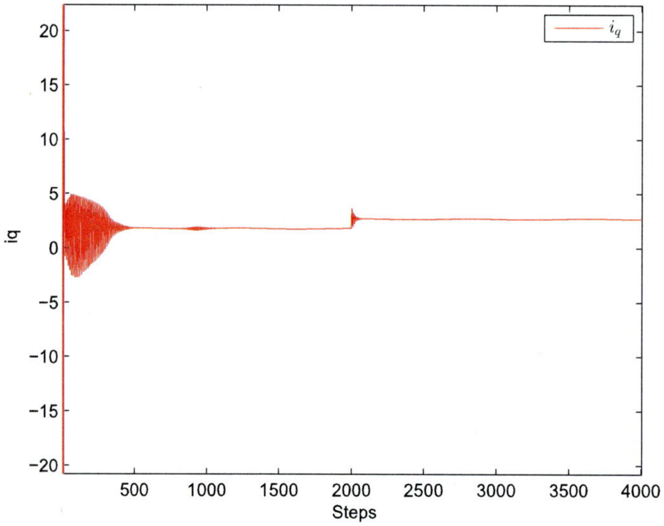

Fig. 4.13 Curve of the i_q without CFC

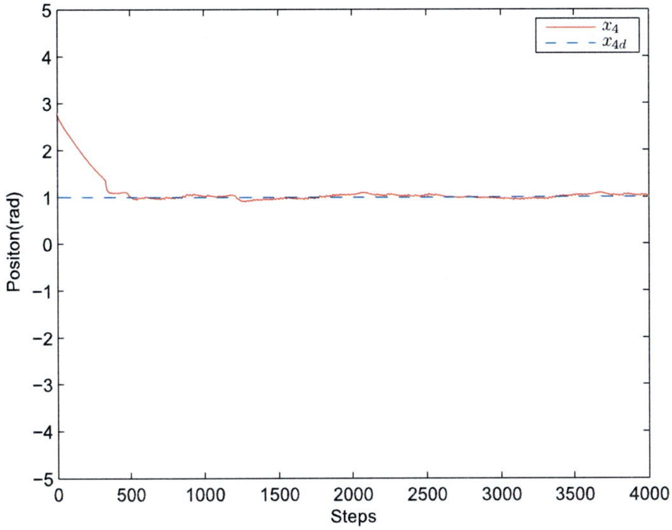

Fig. 4.14 Trajectories of the x_4 and x_{4d} for CFC

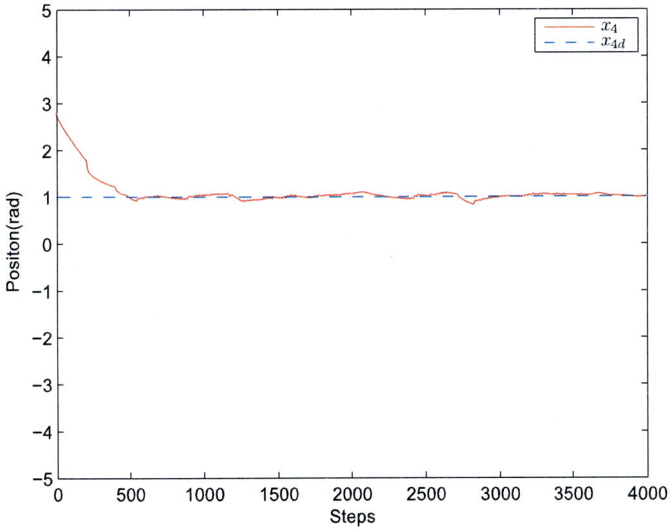

Fig. 4.15 Trajectories of the x_4 and x_{4d} without CFC

4.5 Conclusion

This chapter proposed a neural networks adaptive discrete-time position tracking control for the induction motor drive system based on the command filtered backstepping technique. The euler method is used to describe the discrete-time dynamic mathematical model of the IM. Moreover, the command filtered method is utilized not only to solve the noncausal problem, but also to overcome the "explosion of complexity" problem emerged in traditional backstepping design. In addition, the NNs are employed to approximate the unknown nonlinear functions and the adaptive NNs controllers ensure the tracking error converges to a small bounded neighborhood of the area. The results of the simulation demonstrate the effectiveness and robustness of the proposed approach.

References

1. Li, P., Chen, J., Cai, T., Wang, G.H.: Adaptive robust dynamic surface control of pure-feedback systems using self-constructing neural networks. Int. J. Innov. Comput. Inf. Control **9**(7), 2839–2860 (2013)
2. Tong, S.C., Li, Y., Li, Y.M., Liu, Y.J.: Observer-based adaptive fuzzy backstepping control for a class of stochastic nonlinear strict-feedback systems. IEEE Trans. **41**(6), 1693–1704 (2011)
3. Yu, H.S., Yu, J.P., Liu, J., Song, Q.: Nonlinear control of induction motors based on state error PCH and energy-shaping principle. Nonlinear Dyn. **72**(1–2), 49–59 (2013)
4. Leu, V.Q., Choi, H.H., Jung, J.W.: Fuzzy sliding mode speed controller for PM synchronous motors with a load torque observer. IEEE Trans. Power Electron. **27**(3), 1530–1539 (2012)
5. Li, H., Gao, H., Shi, P., Zhao, X.: Fault-tolerant control of Markovian jump stochastic systems via the augmented sliding mode observer approach. Automatica **50**(7), 1825–1834 (2014)
6. Li, H.Y., Yu, J.Y., Hilton, C., Liu, H.H.: Adaptive sliding mode control for nonlinear active suspension vehicle systems using T-S fuzzy approach. IEEE Trans. Ind. Electron. **60**(8), 3328–3338 (2013)
7. Wang, Y.C., Cao, L.J., Zhang, S.X., Hu, X.X., Yu, F.X.: Command filtered adaptive fuzzy backstepping control method of uncertain nonlinear systems. IET Contr. Theory Appl. **10**(10), 1134–1141 (2016)
8. Liu, Y.J., Wen, G.X., Tong, S.C.: Direct adaptive NN control for a class of discrete-time nonlinear strict-feedback systems. Neurocomputing **73**(13), 2498–2505 (2010)
9. Liu, H.L., Shi, X.P., Bi, X.T., Zhang, J.: Backstepping based terminal sliding mode control for rendezvous and docking with a tumbling spacecraft. Int. J. Innov. Comput. Inf. Control **12**(3), 929–940 (2016)
10. Yu, J.P., Shi, P., Yu, H.S., Chen, B., Lin, C.: Approximation-based discrete-time adaptive position tracking control for interior permanent magnet synchronous motors. IEEE Trans. Cybern. **45**(7), 1363–1371 (2015)
11. Yu, J.P., Shi, P., Dong, W., Lin, C.: Adaptive fuzzy control of nonlinear systems with unknown dead zones based on command filtering. IEEE Trans. Fuzzy Syst. **26**(1), 46–55 (2018)
12. Yu, J.P., Ma, Y.M., Yu, H.S., Lin, C.: Reduced-order observer-based adaptive fuzzy tracking control for chaotic permanent magnet synchronous motors. Neurocomputing **214**, 201–209 (2016)
13. Yu, J.P., Ma, Y.M., Yu, H.S., Lin, C.: Adaptive fuzzy dynamic surface control for induction motors with iron losses in electric vehicle drive systems via backstepping. Inf. Sci. **376**, 172–189 (2017)

14. Yu, J.P., Shi, P., Dong, W., Chen, B., Lin, C.: Neural network-based adaptive dynamic surface control for permanent magnet synchronous motors. IEEE Trans. Neural Netw. Learn. Syst. **26**(3), 640–645 (2015)
15. Yu, J.P., Chen, B., Yu, H.S., Lin, C., Ji, Z.J., Cheng, X.Q.: Position tracking control for chaotic permanent magnet synchronous motors via indirect adaptive neural approximation. Neurocomputing **156**, 245–251 (2015)
16. Zhang, Y., Akujuobi, C.M., Ali, W.H., Tolliver, C.L., Shieh, L.S.: Load disturbance resistance speed controller design for PMSM. IEEE Trans. Ind. Electron. **53**(4), 1198–1208 (2006)
17. Chen, M., Tao, G., Jiang, B.: Dynamic surface control using neural networks for a class of uncertain nonlinear systems with input saturation. IEEE Trans. Neural Netw. Learn. Syst. **26**(9), 2086–2097 (2015)
18. Chen, M., Ge, S.S., How, B.: Robust adaptive neural network control for a class of uncertain MIMO nonlinear systems with input nonlinearities. IEEE Trans. Neural Netw. **21**(5), 796–812 (2010)
19. Ma, J.J., Zheng, Z.Q., Li, P.: Adaptive dynamic surface control of a class of nonlinear systems with unknown direction control gains and input saturation. IEEE Trans. Cybern. **45**(4), 728–741 (2015)
20. Hua, C.C., Wang, Q.G., Guan, X.P.: Adaptive fuzzy outputfeedback controller design for nonlinear time-delay systems with unknown control direction. IEEE Trans. Syst. Man Cybern. Part B Cybern. **39**(2), 363–374 (2009)
21. Liu, Y.J., Tong, S.: Adaptive NN tracking control of uncertain nonlinear discrete-time systems with nonaffine Dead-Zone input. IEEE Trans. Cybern. **45**(3), 497–505 (2017)

Chapter 5
Adaptive Fuzzy Control for IM Stochastic Nonlinear Systems Based on CFC

In this chapter, an adaptive fuzzy control method based on the command filtered is proposed to solve the problems of stochastic disturbance and input saturation of the IM drive system. Firstly, the fuzzy logic system (FLS) is employed to cope with the stochastic nonlinear functions in IM drive systems. Secondly, the adaptive backstepping method is used to design controllers and the quartic Lyapunov function is selected as the stochastic Lyapunov function. Then the command filtered control (CFC) technology is utilized to deal with the "explosion of complexity" in conventional backstepping, and the compensation signal is designed to eliminate the filtering error. Finally, the simulation results verify the effectiveness and superiority of the proposed method.

5.1 Introduction

Stochastic disturbances are always regarded as the common sources of the instability of IM in the actual industrial environment, for example, the voltage has stochastic surges and the external load is randomly switched [1–5]. Moreover, the damping torque, the torsional elastic torque and the magnetic saturation can make some IM parameter variables to a certain extent [6–8], for instance, self-inductance, mutual inductance, winding resistance and so on. The IM dynamic response and control accuracy will be influenced by these stochastic disturbances. In addition, the input saturation is also a common constraint [9, 10], which may make the control less effective and even damage the stability of the system. Therefore, it is significance to study input saturation and stochastic disturbance problems to improve the performance of IM drive systems.

In this chapter, an adaptive fuzzy control method based on command filtered is proposed for IM stochastic nonlinear systems with input saturation. The innovations

J. Yu et al., *Intelligent Backstepping Control for the Alternating-Current Drive Systems*, Studies in Systems, Decision and Control 349,
https://doi.org/10.1007/978-3-030-67723-7_5

of this chapter are summarized as follows: (1) the FLS is utilized to approximate unknown stochastic nonlinear functions of IM drive systems during the controller design process, which makes the proposed method more suitable to the practical industrial environment, (2) stochastic disturbances are considered in this chapter, which enhances the robustness and stability of the system, (3) the CFC method is utilized to solve the problem of "explosion of complexity", and filtering error is reduced by the designed compensating signal, which overcomes the shortcoming of the DSC method, and achieves better performance and higher control precision of IM.

5.2 The IM Drive Systems Mathematical Model

The IMs stochastic systems model [11] can be described as the following form

$$\begin{cases} dx_1 = \left(\frac{a_1}{J}x_2x_3 - \frac{T_L}{J}\right)dt + \psi_1^T d\varphi, \\ dx_2 = \left(b_1x_2 + b_2x_1x_3 - b_3x_1x_4 - b_4\frac{x_2x_4}{x_3} + b_5u_q\right)dt + \psi_2^T d\varphi, \\ dx_3 = (c_1x_3 + b_4x_4)dt + \psi_3^T d\varphi, \\ dx_4 = \left(b_1x_4 + d_2x_3 + b_3x_1x_2 + b_4\frac{x_2^2}{x_3} + b_5u_d\right)dt + \psi_4^T d\varphi. \end{cases} \tag{5.1}$$

For stochastic control system $dx = p(x)dt + g(x)d\varphi$, where φ denotes r-dimensional standard Brownian motion, $p(\cdot)$ and $g(\cdot)$ satisfy lipschitz condition and linear growth condition, and $p(0) = 0$, $g(0) = 0$. The following concepts are proposed:

Definition 5.1 [12] For any given $V(x)$, define the stochastic differential operator $\mathscr{L}$ as

$$\mathscr{L}V = \frac{\partial V}{\partial x}p + \frac{1}{2}\mathrm{Tr}\left\{g^T\frac{\partial^2 V}{\partial x^2}g\right\}, \tag{5.2}$$

where $\mathrm{Tr}(\cdot)$ denotes the trace of a matrix.

In this chapter, taking the saturation restriction into consideration, the control input u can be described as

$$u = \mathrm{sat}(W) = \begin{cases} \mathrm{sign}(W)u_{\max}, |W| \geq u_{\max} \\ W, |W| < u_{\max} \end{cases}, \tag{5.3}$$

where $u_{\max}$ is an unknown parameter. It's remarkable that when $|W| = u_{\max}$, there is a sharp corner. Therefore, a smooth function is utilized to solve this problem.

$$c\left(W\right) = u_{\max} * \tanh\left(W/u_{\max}\right) = u_{\max} * \frac{e^{W/u_{\max}} - e^{-W/u_{\max}}}{e^{W/u_{\max}} + e^{-W/u_{\max}}}, \tag{5.4}$$

then we can get

$$u = \text{sat}(W) = c(W) + d(W), \tag{5.5}$$

where $d\left(W\right) = \text{sat}\left(W\right) - c\left(W\right)$ is a bounded function,

$$\left|d\left(W\right)\right| = \left|\text{sat}\left(W\right) - c\left(W\right)\right| \leq u_{\max}\left(1 - \tanh\left(1\right)\right) = D. \tag{5.6}$$

Furthermore, it can be proved by the mean-value theorem that there exits a constant η with $0 < \eta < 1$, such that

$$c\left(W\right) = c\left(W_0\right) + c_{W_\eta}\left(W - W_0\right), \tag{5.7}$$

where $c_{W_\eta} = \frac{\partial c(W)}{\partial W}|_{W=W_\eta}$, $W_\eta = \eta W + \left(1 - \eta\right) W_0$. By choosing $W_0 = 0$, the above equation can be rewritten as

$$c\left(W\right) = c_{W_\eta} W. \tag{5.8}$$

Then we can obtain

$$u = c_{W_\eta} W + d\left(W\right), \tag{5.9}$$

and there exists a positive constant c_m, which satisfies $0 < c_m < c_{W_\eta} \leq 1$.

Lemma 5.2 *[13] If there exists a stochastic Lyapunov function $V : R^n \to R$ which is positive definite, radially unbounded, and twice continuously differentiable, and constants $c_0 > 0$, $d_0 \geq 0$ such that*

$$\mathscr{L}V\left(x\right) \leq -c_0 V\left(x\right) + d_0, \tag{5.10}$$

then the stochastic system has a unique solution almost surely and all the signals in the closed loop system are bounded in probability.

Lemma 5.3 *[14] For a continuous function $f(Z)$ defined on a compact set Ω_Z, then for any $\varepsilon > 0$, there exists a FLS $W^T S(Z)$ such that*

$$f(Z) = W^T S(Z) + \delta(Z), \tag{5.11}$$

for all $Z \in \Omega_Z$, $\delta(Z)$ is approximation error and satisfies $|\delta(Z)| \leq \varepsilon$.

Lemma 5.4 *[15] The form of command filter is constructed as follows:*

$$\begin{aligned}\dot{\varpi}_1 &= \omega_n \varpi_2,\\ \dot{\varpi}_2 &= -2\zeta\omega_n\varpi_2 - \omega_n\left(\varpi_1 - \beta_1\right),\end{aligned} \tag{5.12}$$

for all $t \geq 0$, *if the input signal* β_1 *satisfies* $|\dot{\beta}_1| \leq q_1$, $|\ddot{\beta}_1| \leq q_2$, *where* q_1, q_2 *are positive constants and* $\varpi_1(0) = \alpha_1(0)$, $\varpi_2(0) = 0$, *then for any* $\mu > 0$, *there exists* ζ, ω_n, *we have* $|\varpi - \beta_1| \leq \mu$, *and* $|\dot{\varpi}_1|$, $|\ddot{\varpi}_1|$, $|\dddot{\varpi}_1|$ *are bounded.*

5.3 Adaptive Fuzzy Control Based on CFC for IM Stochastic Nonlinear Systems

In this section, an adaptive fuzzy controller for IM stochastic nonlinear systems with input saturation based on command filtered will be proposed.

For simplicity, define unknown constants θ_i

$$\theta_1 = \|W_1\|^2, \theta_2 = \frac{1}{b_q}\|W_2\|^2, \theta_3 = \|W_3\|^2, \theta_4 = \frac{1}{b_d}\|W_4\|^2, \tag{5.13}$$

where $\hat{\theta}_i$ is the estimation of θ_i, $\tilde{\theta}_i = \theta_i - \hat{\theta}_i$, and b_q, b_d are given later.

Step 1: The tracking error is defined as $z_1 = x_1 - x_{1d}$, and the compensating signal is $v_1 = z_1 - \xi_1$. Consider the stochastic Lyapunov function candidate as $V_1 = J\frac{v_1^4}{4} + \frac{1}{2\pi_1}\tilde{\theta}_1^2$, we can get

$$\mathscr{L}V_1 = v_1^3(a_1x_2x_3 - T_L - J\dot{x}_{1d} - J\dot{\xi}_1) + \frac{3}{2}v_1^2\psi_1^T\psi_1 - \frac{1}{\pi_1}\tilde{\theta}_1\dot{\hat{\theta}}_1. \tag{5.14}$$

Remark 5.5 In the real system, the load T_L will be bounded, so we assume there exists its upper $d > 0$, which means that $|T_L| \leq d$. ♦

Obviously,

$$-v_1^3T_L \leq \frac{3}{4}v_1^4 + \frac{1}{4}d^4, \tag{5.15}$$

$$\frac{3}{2}v_1^2\psi_1^T\psi_1 \leq \frac{3}{4}l_1^{-2}v_1^4\|\psi_1\|^4 + \frac{3}{4}l_1^2, \tag{5.16}$$

where $l_i > 0 (i = 1, 2, 3, 4)$ are designed constants, let $f_1(Z_1) = a_1x_2x_3 - x_2 + \frac{3}{4}l_1^{-2}v_1\|\psi_1\|^4$, $Z_1 = [x_1, x_2, x_3, \xi_1, x_{1d}, \dot{x}_{1d}] \in R^6$, and according to Lemma 5.3, there exists a FLS $W_1^TS_1(Z_1)$ such that $f_1(Z_1) = W_1^TS_1(Z_1) + \delta_1(Z_1)$, $|\delta_1(Z_1)| \leq \varepsilon_1$.

Furthermore, by using Young's inequality, one has

$$v_1^3f_1(Z_1) \leq \frac{1}{2j_1^2}v_1^6\theta_1S_1^TS_1 + \frac{1}{2}j_1^2 + \frac{3}{4}v_1^4 + \frac{1}{4}\varepsilon_1^4. \tag{5.17}$$

Construct virtual control law α_1, compensating signal ξ_1 as

$$\alpha_1 = -k_1 z_1 - \frac{3}{2}v_1 - \frac{1}{2j_1^2}v_1^3\hat{\theta}S_1^T S_1 + \hat{J}\dot{x}_{1d}, \tag{5.18}$$

$$\dot{\xi}_1 = \frac{1}{J}(-k_1\xi_1 + \xi_2 + (x_{1,c} - \alpha_1)). \tag{5.19}$$

Similarly, define $z_2 = x_2 - x_{1,c}$, and $v_2 = z_2 - \xi_2$. Substituting Eqs. (5.15)–(5.19) into Eq. (5.14) gets

$$\begin{aligned}\mathscr{L}V_1 \le &- k_1 v_1^4 + v_1^3 v_2 + \frac{1}{2}j_1^2 + \frac{1}{4}\varepsilon_1^4 + \frac{3}{4}l_1^2 + \frac{1}{4}d^4 \\ &+ \frac{1}{\pi_1}\tilde{\theta}_1(\frac{\pi_1}{2j_1^2}v_1^6 S_1^T S_1 - \dot{\hat{\theta}}_1) + v_1^3(\hat{J} - J)\dot{x}_{1d}.\end{aligned} \tag{5.20}$$

The adaptive law $\hat{\theta}_1$ is chosen as

$$\dot{\hat{\theta}}_1 = \frac{\pi_1}{2j_1^2}v_1^6 S_1^T S_1 - m_1\hat{\theta}_1, \tag{5.21}$$

where $j_i, m_i (i = 1, 2, 3, 4)$ are design parameters.

By combining (5.20) and (5.21), one has

$$\begin{aligned}\mathscr{L}V_1 \le &- k_1 v_1^4 + v_1^3 v_2 + \frac{1}{2}j_1^2 + \frac{1}{4}\varepsilon_1^4 + \frac{1}{4}d^4 + \frac{3}{4}l_1^2 \\ &+ \frac{m_1}{\pi_1}\tilde{\theta}_1\hat{\theta}_1 + v_1^3(\hat{J} - J)\dot{x}_{1d}.\end{aligned} \tag{5.22}$$

Step 2: The compensating signal is $v_2 = z_2 - \xi_2$, with $\dot{\xi}_2 = 0$. Now choose the stochastic Lyapunov function as $V_2 = V_1 + \frac{1}{4}v_2^4 + \frac{b_q}{2\pi_2}\tilde{\theta}_2^2$. Similarly, we can get

$$\mathscr{L}V_2 \le \mathscr{L}V_1 + v_2^3(f_2(Z) + b_5 u_q - \dot{\xi}_2) + \frac{3}{4}l_2^2 - v_2^4 - \frac{b_q}{\pi_2}\tilde{\theta}_2\dot{\hat{\theta}}_2. \tag{5.23}$$

Let $f_2(Z_2) = b_1 x_2 + b_2 x_1 x_3 - b_3 x_1 x_4 - b_4\frac{x_2 x_4}{x_3} - \mathscr{L}x_{1,c} + \frac{3}{4}l_2^{-2}v_2\,\|\psi_2\|^4 + v_2$, $Z_2 = [x_1, x_2, x_3, \xi_1, \hat{\theta}_1, x_{1d}, \dot{x}_{1d}, \ddot{x}_{1d}] \in R^8$, where

$$\begin{aligned}\mathscr{L}x_{1,c} = &\frac{\partial x_{1,c}}{\partial x_1}(\frac{a_1}{J}x_2 x_3 - \frac{T_L}{J}) + \frac{\partial x_{1,c}}{\partial \xi_1}\dot{\xi}_1 + \frac{\partial x_{1,c}}{\partial \hat{\theta}_1}\dot{\hat{\theta}}_1 \\ &+ \frac{1}{2}\sum_{j=0}^{1}\frac{\partial x_{1,c}}{\partial x_{1d}^j}x_{1d}^{(j+1)} + \frac{1}{2}\frac{\partial^2 x_{1,c}}{\partial^2 x_1}\psi_1^T\psi_1.\end{aligned} \tag{5.24}$$

Similar to Step 1, one has

$$\upsilon_2^3 f_2(Z_2) \leq \frac{1}{2j_2^2}\upsilon_2^6 \|W_2\|^2 S_2^T S_2 + \frac{1}{2}j_2^2 + \frac{3}{4}\upsilon_2^4 + \frac{1}{4}\varepsilon_2^4, \tag{5.25}$$

$$\upsilon_1^3 \upsilon_2 \leq \frac{3}{4}\upsilon_1^4 + \frac{1}{4}\upsilon_2^4. \tag{5.26}$$

Let $u_q = c_{\nu_{q\eta}}\nu_q + d(\nu_q)$, and $\left|d\left(\nu_q\right)\right| \leq D_q$, then we can obtain υ_2^3 $b_5 u_q = \upsilon_2^3 b_5 c_{\nu_{q\eta}}\nu_q + \upsilon_2^3 b_5 d(\nu_q)$. From $0 < c_m < c_{q\nu_\eta} \leq 1$, there exists a positive number b_q that $b_5 c_{\nu_{q\eta}} \geq b_q$.

Design the control law ν_q as $\nu_q = -k_2 z_2 - \frac{1}{2j_2^2}\upsilon_2^3\hat{\theta}S_2^T S_2$ and we have

$$\upsilon_2^3 b_5 c_{\nu_{q\eta}}\nu_q \leq -k_2 b_q \upsilon_2^4 - \frac{1}{2j_2^2} b_q \upsilon_2^6 \hat{\theta} S_2^T S_2, \tag{5.27}$$

$$\upsilon_2^3 b_5 d(\nu_q) \leq \frac{3}{4}\upsilon_2^4 + \frac{1}{4}b_5^4 D_q^4. \tag{5.28}$$

Furthermore, substituting Eqs. (5.25)–(5.28) into Eq. (5.23), yields

$$\begin{aligned}\mathscr{L}V_2 \leq& -(k_1 - \frac{3}{4})\upsilon_1^4 - (k_2 b_q - \frac{3}{4})\upsilon_2^4 + \frac{1}{4}d^4 + \frac{1}{4}b_5^4 D_q^4 + \frac{1}{2}\sum_{i=1}^{2} j_i^2 \\ &+ \frac{1}{4}\sum_{i=1}^{2}\varepsilon_i^4 + \frac{3}{4}\sum_{i=1}^{2} l_i^2 + \upsilon_1^3(\hat{J} - J)\dot{x}_{1d} \\ &+ \frac{m_1}{\pi_1}\tilde{\theta}_1\hat{\theta}_1 + \frac{b_q}{\pi_2}\tilde{\theta}_2(\frac{\pi_2}{2j_2^2}\upsilon_2^6 S_2^T S_2 - \dot{\hat{\theta}}_2).\end{aligned} \tag{5.29}$$

Define an adaptive law $\hat{\theta}_2$ as

$$\dot{\hat{\theta}}_2 = \frac{\pi_2}{2j_2^2}\upsilon_2^6 S_2^T S_2 - m_2\hat{\theta}_2. \tag{5.30}$$

Similarly, it follows

$$\begin{aligned}\mathscr{L}V_2 \leq& -(k_1 - \frac{3}{4})\upsilon_1^4 - (k_2 b_q - \frac{3}{4})\upsilon_2^4 + \frac{1}{2}\sum_{i=1}^{2} j_i^2 + \frac{1}{4}\sum_{i=1}^{2}\varepsilon_i^4 + \frac{3}{4}\sum_{i=1}^{2} l_i^2 \\ &+ \frac{1}{4}d^4 + \frac{1}{4}b_5^4 D_q^4 + \upsilon_1^3(\hat{J} - J)\dot{x}_{1d} + \frac{m_1}{\pi_1}\tilde{\theta}_1\hat{\theta}_1 + \frac{b_q m_2}{\pi_2}\tilde{\theta}_2\hat{\theta}_2.\end{aligned} \tag{5.31}$$

Step 3: The tracking error is defined as $z_3 = x_3 - x_{3d}$, and $\upsilon_3 = z_3 - \xi_3$, choose the stochastic Lyapunov function candidate as $V_3 = V_2 + \frac{1}{4}\upsilon_3^4 + \frac{1}{2\pi_3}\tilde{\theta}_3^2$. Similarly, it follows

$$\mathscr{L}V_3 \leq \mathscr{L}V_2 + \upsilon_3^3(c_1 x_3 + b_4 x_4 - \dot{x_{3d}} - \dot{\xi}_3) + \frac{3}{2}\upsilon_3^2\psi_3^T\psi_3 - \frac{1}{\pi_3}\tilde{\theta}_3\dot{\hat{\theta}}_3. \tag{5.32}$$

Similarly, define that $f_3(Z_3) = c_1x_3 + b_4x_4 - \dot{x}_{3d} - x_4 + \frac{3}{4}v_3 + \frac{3}{4}l_3^{-2}v_3\|\psi_3\|^4$, where $Z_3 = [x_1, x_2, x_3, x_4, \xi_3, x_{3d}, \dot{x}_{3d}] \in R^7$ and construct α_3, ξ_3 and θ_3 as

$$\dot{\xi}_3 = -k_3\xi_3 + \xi_4 + (x_{3,c} - \alpha_3), \tag{5.33}$$

$$\alpha_3 = -k_3z_3 - \frac{1}{2j_3^2}v_3^3\hat{\theta}S_3^T S_3, \tag{5.34}$$

$$\dot{\hat{\theta}}_3 = \frac{\pi_3}{2j_3^2}v_3^6 S_3^T S_3 - m_3\hat{\theta}_3. \tag{5.35}$$

Similarly to Eq. (5.22), one has

$$\begin{aligned}\mathscr{L}V_3 \le& -(k_1 - \frac{3}{4})v_1^4 - (k_2b_q - \frac{3}{4})v_2^4 - k_3v_3^4 + \frac{1}{2}\sum_{i=1}^{3} j_i^2 + \frac{1}{4}\sum_{i=1}^{3}\varepsilon_i^4 + \frac{3}{4}\sum_{i=1}^{3} l_i^2\\ &+\frac{1}{4}d^4 + \frac{1}{4}b_5^4D_q^4 + v_3^3v_4 + v_1^3(\hat{J} - J)\dot{x}_{1d}\\ &+\frac{m_1}{\pi_1}\tilde{\theta}_1\hat{\theta}_1 + \frac{b_qm_2}{\pi_2}\tilde{\theta}_2\hat{\theta}_2 + \frac{m_3}{\pi_3}\tilde{\theta}_3\hat{\theta}_3.\end{aligned} \tag{5.36}$$

Step 4: At this step, we will construct the control law u_d. Choose $z_4 = x_4 - x_{3,c}$, $v_4 = z_4 - \xi_4$, with $\dot{\xi}_4 = 0$. Let $u_d = c_{\nu_{d\eta}}\nu_d + d(\nu_d)$, and from $0 < c_m < c_{\nu_{d\eta}} \le 1$, there exists a positive number b_d that $b_5c_{\nu_{d\eta}} \ge b_d$. Choose the stochastic Lyapunov function candidate as $V_4 = V_3 + \frac{1}{4}v_4^4 + \frac{b_d}{2\pi_4}\tilde{\theta}_4^2$ and it arrives

$$\mathscr{L}V_4 \le \mathscr{L}V_3 + v_4^3(f_4 + b_5u_d - \dot{\xi}_4) + \frac{3}{4}l_4^2 - v_4^4 - \frac{b_d}{\pi_4}\tilde{\theta}_4\dot{\hat{\theta}}_4, \tag{5.37}$$

where $f_4(Z_4) = b_1x_4 + d_2x_3 + b_3x_1x_2 + b_4\frac{x_2^2}{x_3} - \mathscr{L}x_{3,c} + \frac{3}{4}l_4^{-2}v_4\|\psi_4\|^4 + v_4$, $Z_4 = [x_1, x_2, x_3, \xi_1, \hat{\theta}_3, x_{3d}, \dot{x}_{3d}, \ddot{x}_{3d}] \in R^8$,

$$\begin{aligned}\mathscr{L}x_{3,c} =& \frac{\partial x_{3,c}}{\partial x_3}(c_1x_3 + b_4x_4) + \frac{\partial x_{3,c}}{\partial \xi_3}\dot{\xi}_3 + \frac{\partial x_{3,c}}{\partial \hat{\theta}_3}\dot{\hat{\theta}}_3\\ &+\frac{1}{2}\sum_{j=0}^{1}\frac{\partial x_{3,c}}{\partial x_{3d}^j}x_{3d}^{(j+1)} + \frac{1}{2}\frac{\partial^2 x_{3,c}}{\partial^2 x_3}\psi_3^T\psi_3.\end{aligned} \tag{5.38}$$

For given $\varepsilon_4 > 0$, we design ν_d as $\nu_d = -k_4z_4 - \frac{1}{2j_4^2}z_4^3\hat{\theta}S_4^T S_4$. Similarly, we can get

$$\begin{cases} v_4^3 f_4(Z_4) \leq \dfrac{1}{2j_4^2} v_4^6 \|W_4\|^2 S_4^T S_4 + \dfrac{1}{2} j_4^2 + \dfrac{3}{4} v_4^4 + \dfrac{1}{4} \varepsilon_4^4, \\ v_3^3 v_4 \leq \dfrac{3}{4} v_3^4 + \dfrac{1}{4} v_4^4, \\ v_4^3 b_5 c_{\nu_{d\eta}} \nu_d \leq -k_4 b_d v_4^4 - \dfrac{1}{2j_4^2} b_d v_4^6 \hat{\theta} S_4^T S_4, \\ v_4^3 b_5 d(\nu_d) \leq \dfrac{3}{4} v_4^4 + \dfrac{1}{4} b_5^4 D_d^4. \end{cases} \tag{5.39}$$

By substituting Eq. (5.39) into (5.37), it can be verified easily that

$$\begin{aligned} \mathscr{L}V_4 \leq & -(k_1 - \frac{3}{4})v_1^4 - (k_2 b_q - \frac{3}{4})v_2^4 - (k_3 - \frac{3}{4})v_3^4 - (k_4 b_d - \frac{3}{4})v_4^4 + \frac{1}{4}d^4 \\ & + \frac{1}{4} b_5^4 D_q^4 + \frac{1}{4} b_5^4 D_d^4 + \frac{1}{2}\sum_{i=1}^{4} j_i^2 + \frac{1}{4}\sum_{i=1}^{4}\varepsilon_i^4 + \frac{3}{4}\sum_{i=1}^{4} l_i^2 + v_1^3(\hat{J} - J)\dot{x}_{1d} \\ & + \frac{m_1}{\pi_1}\tilde{\theta}_1\hat{\theta}_1 + \frac{b_q m_2}{\pi_2}\tilde{\theta}_2\hat{\theta}_2 + \frac{m_3}{\pi_3}\tilde{\theta}_3\hat{\theta}_3 + \frac{b_d}{\pi_4}\tilde{\theta}_4(\frac{\pi_4}{2j_4^2} v_4^6 S_4^T S_4 - \dot{\hat{\theta}}_4). \end{aligned} \tag{5.40}$$

The adaptive laws are constructed as

$$\begin{aligned} \dot{\hat{\theta}}_4 &= \frac{\pi_4}{2j_4^2} v_4^6 S_4^T S_4 - m_4 \hat{\theta}_4, \\ \dot{\hat{J}} &= \lambda_1 v_1^3 \dot{x}_{1d} + \lambda_2 \hat{J}, \end{aligned} \tag{5.41}$$

where $\tilde{J} = J - \hat{J}$, and $\lambda_1, \lambda_2, \pi_i, m_i$ $(i = 1, 2, 3, 4)$ are positive design parameters.

And similar to Eq. (5.31), we can obtain

$$\begin{aligned} \mathscr{L}V_4 \leq & -(k_1 - \frac{3}{4})v_1^4 - (k_2 b_q - \frac{3}{4})v_2^4 - (k_3 - \frac{3}{4})v_3^4 - (k_4 b_d - \frac{3}{4})v_4^4 + \frac{1}{4}d^4 \\ & + \frac{1}{4} b_5^4 D_q^4 + \frac{1}{4} b_5^4 D_d^4 + \frac{1}{2}\sum_{i=1}^{4} j_i^2 + \frac{1}{4}\sum_{i=1}^{4}\varepsilon_i^4 + \frac{3}{4}\sum_{i=1}^{4} l_i^2 - v_1^3 \tilde{J}\dot{x}_{1d} \\ & + \frac{m_1}{\pi_1}\tilde{\theta}_1\hat{\theta}_1 + \frac{b_q m_2}{\pi_2}\tilde{\theta}_2\hat{\theta}_2 + \frac{m_3}{\pi_3}\tilde{\theta}_3\hat{\theta}_3 + \frac{b_d m_4}{\pi_4}\tilde{\theta}_4\hat{\theta}_4. \end{aligned} \tag{5.42}$$

Theorem 5.6 *Consider system (5.1) and reference signal x_{1d} and x_{3d}. If the virtual control signals are constructed as in (5.18) and (5.34), the adaptive law is designed as in (5.21), (5.30), (5.35) and (5.41), then we choose the adaptive fuzzy controllers ν_d and ν_q such that the resulting tracking errors converge to the origin's small neighborhood. Also, all closed-loop signals of the controlled system are bounded.*

Proof To address the stability analysis of the resulting closed-loop system, choose the stochastic Lyapunov function candidate as $V = V_4 + \frac{1}{2\lambda_1}\tilde{J}^2$ and it gives

$$\mathscr{L}V \leq \sum_{i=1}^{4}\left(-\bar{k}_i v_i^4 + \frac{1}{2}j_i^2 + \frac{3}{4}l_i^2 + \frac{1}{4}\varepsilon_i^2\right) + \frac{1}{4}d^4 + \frac{1}{4}b_5^4 D_q^4 + \frac{1}{4}b_5^4 D_d^4$$
$$+ \frac{m_1}{\pi_1}\tilde{\theta}_1\hat{\theta}_1 + \frac{b_q m_2}{\pi_2}\tilde{\theta}_2\hat{\theta}_2 + \frac{m_3}{\pi_3}\tilde{\theta}_3\hat{\theta}_3 + \frac{b_d m_4}{\pi_4}\tilde{\theta}_4\hat{\theta}_4 + \frac{\lambda_2}{\lambda_1}\tilde{J}\hat{J}, \tag{5.43}$$

where $\bar{k}_i (i = 1, 2, 3, 4)$ is defined as

$$\bar{k}_1 = k_1 - \frac{3}{4}, \bar{k}_2 = k_2 b_q - \frac{3}{4}, \bar{k}_3 = k_1 - \frac{3}{4}, \bar{k}_4 = k_4 b_d - \frac{3}{4}. \tag{5.44}$$

Obviously, we can get

$$\tilde{\theta}_i\hat{\theta}_i \leq -\frac{1}{2}\tilde{\theta}_i^2 + \frac{1}{2}\theta_i^2, i = 1, 2, 3, 4,$$
$$\tilde{J}\hat{J} \leq -\frac{1}{2}\tilde{J}^2 + \frac{1}{2}J^2. \tag{5.45}$$

Furthermore, we can obtain

$$\mathscr{L}V \leq -\sum_{i=1}^{4}\bar{k}_i v_i^4 - \frac{m_1}{\pi_1}\tilde{\theta}_1^2 - \frac{b_q m_2}{\pi_2}\tilde{\theta}_2^2 - \frac{m_3}{\pi_3}\tilde{\theta}_3^2 - \frac{b_d m_4}{\pi_4}\tilde{\theta}_4^2 - \frac{\lambda_2}{\lambda_1}\tilde{J}^2$$
$$+ \sum_{i=1}^{4}\left(\frac{1}{2}j_i^2 + \frac{3}{4}l_i^2 + \frac{1}{4}\varepsilon_i^2\right) + \frac{1}{4}d^4 + \frac{1}{4}b_5^4 D_q^4 + \frac{1}{4}b_5^4 D_d^4$$
$$+ \frac{m_1}{\pi_1}\theta_1^2 + \frac{b_q m_2}{\pi_2}\theta_2^2 + \frac{m_3}{\pi_3}\theta_3^2 + + \frac{b_d m_4}{\pi_4}\theta_4^2 + \frac{\lambda_2}{\lambda_1}J^2. \tag{5.46}$$

Next, let

$$c_0 = \min\left\{4\bar{k}_1/J, 4\bar{k}_i, 2m_1, 2m_2, 2m_3, 2m_4, 2\lambda_2, i = 2, 3, 4\right\},$$
$$d_0 = \sum_{i=1}^{4}\left(\frac{1}{2}j_i^2 + \frac{3}{4}l_i^2 + \frac{1}{4}\varepsilon_i^2\right) + \frac{1}{4}d^4 + \frac{1}{4}b_5^4 D_q^4 + \frac{1}{4}b_5^4 D_d^4$$
$$+\frac{m_1}{\pi_1}\theta_1^2 + \frac{b_q m_2}{\pi_2}\theta_2^2 + \frac{m_3}{\pi_3}\theta_3^2 + \frac{b_d m_4}{\pi_4}\theta_4^2 + \frac{\lambda_2}{\lambda_1}J^2.$$

The Eq. (5.46) can be summed as the following form

$$\mathscr{L}V \leq -c_0 V + d_0. \tag{5.47}$$

Therefore, $\tilde{J}$, $\tilde{\theta}_i$, α_i, and v_i are bounded in probability. Based on Lemma 5.2, it implies that

$$\mathscr{E}\left[V(t)\right] \leq e^{-c_0 t}\left[V(0)\right] + \frac{d_0}{c_0}, \forall t > 0. \tag{5.48}$$

From the Eq. (5.48), it follows that

$$\mathscr{E}\left[|v_i|^4\right] \leq 4\mathscr{E}\left[V(t)\right], i = 1, 2, 3, 4. \tag{5.49}$$

Remark 5.7 It has been confirmed in [16] that $\|\xi_i\|$ are bounded, $z_i = v_i + \xi_i$, we can have that z_i are bounded. Thus, we may adjust the parameters to guarantee that the tracking error can converge to a small neighborhood around the origin in the sense of mean quartic value. ♦

■

5.4 Simulation Results

To verify the method's effectiveness, parameters of the IM are selected as: $L_s = L_r = 0.0699\,\text{H}$, $L_m = 0.068\,\text{H}$, $R_s = 0.1\,\Omega$, $R_r = 0.15\,\Omega$, $J = 0.0586\,\text{Kgm}^2$, $n_p = 1$. Choosing the reference signals as $x_{1d} = \begin{cases} 50, 0 \leq t < 10, \\ 60, t \geq 10. \end{cases}$, $x_{3d} = 0.5$, load torque $T_L = 1.0$, and the initial condition is $[0, 0, 0.5, 0]$.

The fuzzy membership functions are given as follows

$$\mu_{F_i^j} = \exp\left[\frac{-(x_i + l)^2}{2}\right],$$

where $i = 1, 2, \ldots, 4$, $j = 1, 2, \ldots, 12, 13$, $l = -6, -5, \ldots, 5, 6$. The fuzzy logic systems contain thirteen nodes, and widths being equal to 1, respectively.

The control parameters are chosen as: $j_i = [5, 5, 5, 5]$, $\bar{k}_i = [60, 100, 200, 120]$, $\pi_i = 4$, $m_i = 0.01 (i = 1, 2, 3, 4)$, $\lambda_1 = 4$, $\lambda_2 = 0.01$, $\zeta = 0.7$, $\omega_n = 400$.

As a comparison, the DSC method is also used to control the IM system, and the control parameters are selected identical with the proposed CFC scheme.

The simulation results are shown in Figs. 5.1, 5.2, 5.3, 5.4, 5.5, 5.6, 5.7, 5.8, 5.9 and 5.10, where Figs. 5.1, 5.2, 5.3, 5.4 and 5.5 show the proposed CFC scheme, and Figs. 5.6, 5.7, 5.8, 5.9 and 5.10 show the DSC technique. Figures 5.1 and 5.6 show the output response x_1 and the reference signal x_{1d}. The tracking errors are displayed in Figs. 5.2 and 5.7. Similarly, the Figs. 5.3 and 5.8 are about the tracking performance of x_3 and x_{3d}. Figures 5.4, 5.9 show the input voltage of u_q, and Figs. 5.5, 5.10 show the input voltage of u_d.

Remark 5.8 It can be seen from Figs. 5.1, 5.2, 5.6 and 5.7 that both the two methods can track the given reference signal well, but the error of CFC is smaller than the DSC method. In addition, the overshoots of our proposed controllers are far less than the DSC method from Figs. 5.4, 5.5, 5.9 and 5.10. ♦

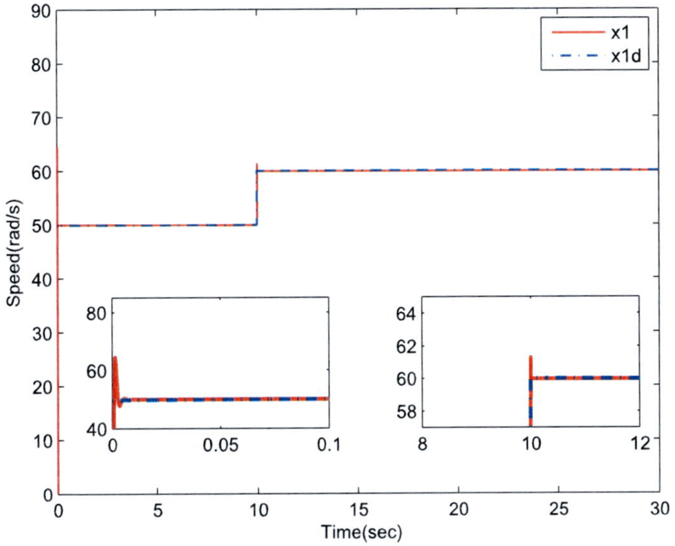

Fig. 5.1 Trajectories of the x_1 and x_{1d} for CFC

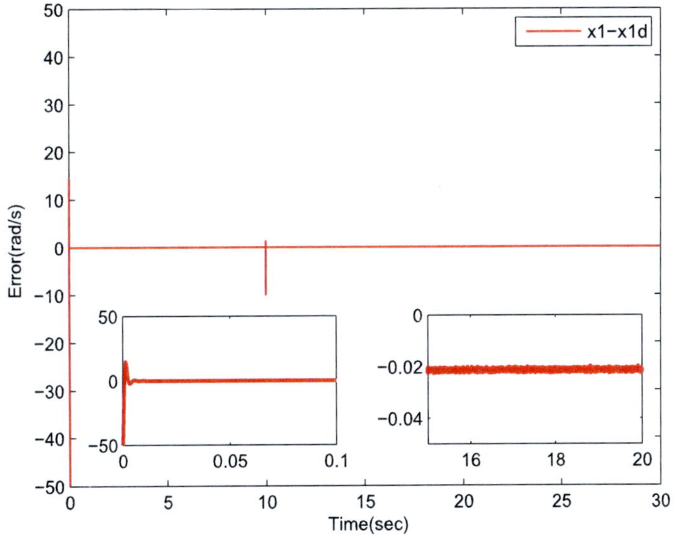

Fig. 5.2 The tracking error of x_1 and x_{1d} for CFC

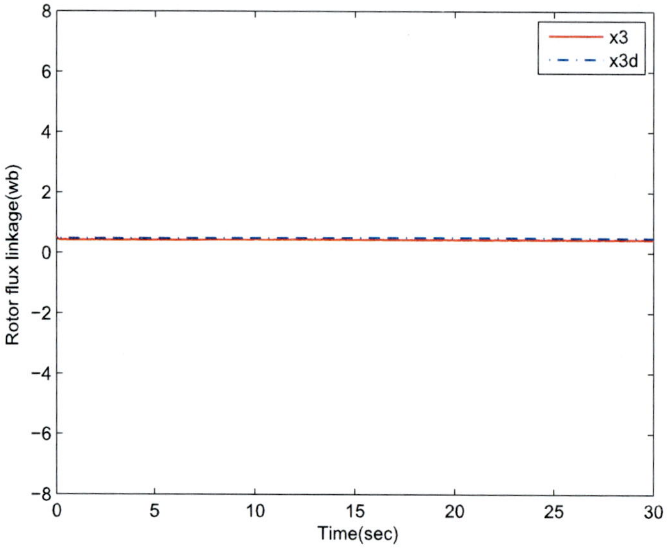

Fig. 5.3 Trajectories of the x_3 and x_{3d} for CFC

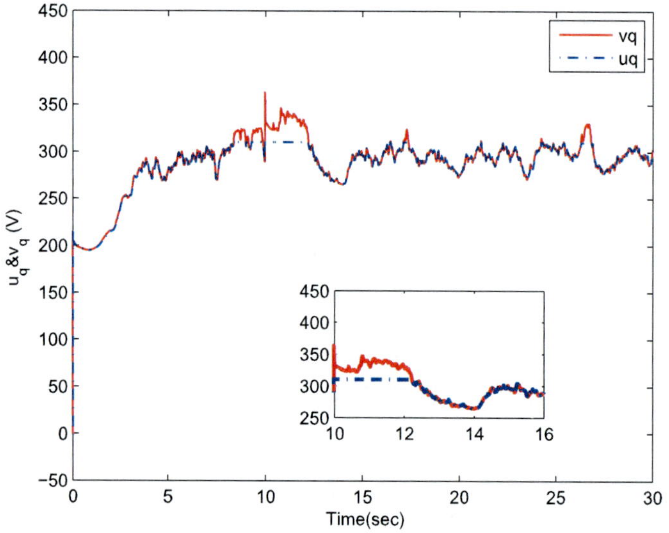

Fig. 5.4 Curves of the u_q and v_q for CFC

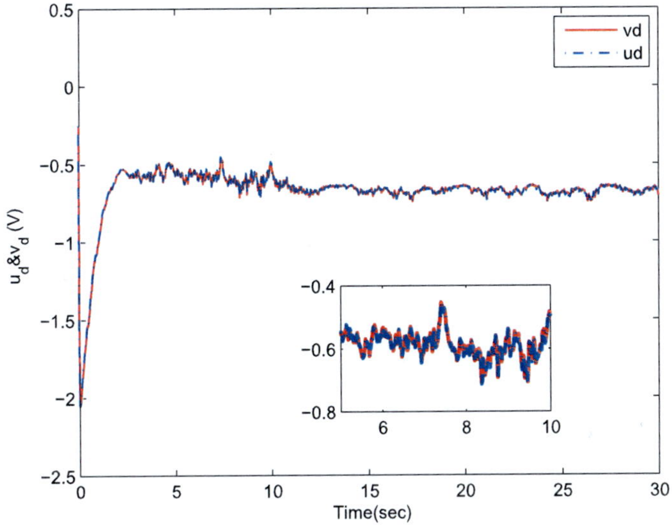

Fig. 5.5 Curves of the u_d and v_d for CFC

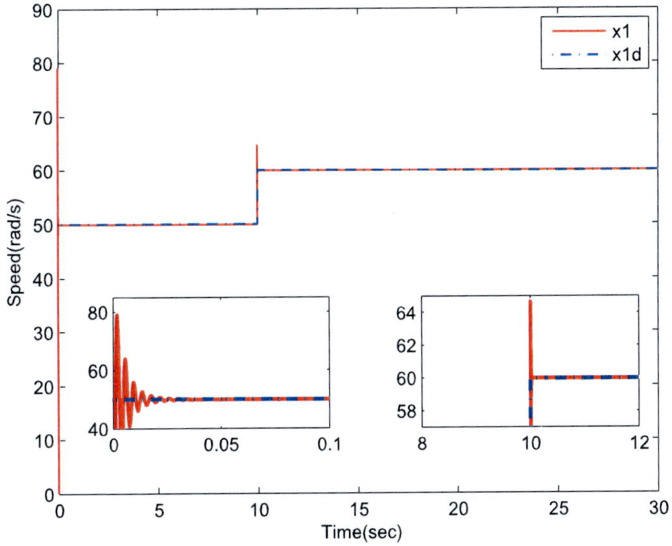

Fig. 5.6 Trajectories of the x_1 and x_{1d} for DSC

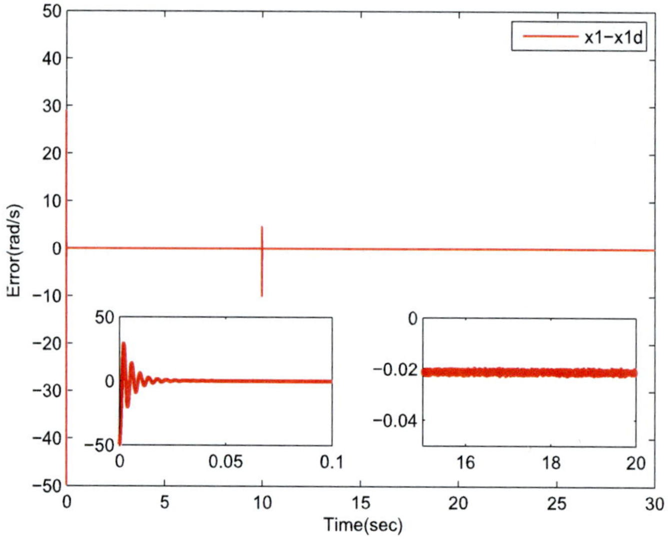

Fig. 5.7 The tracking error of x_1 and x_{1d} for DSC

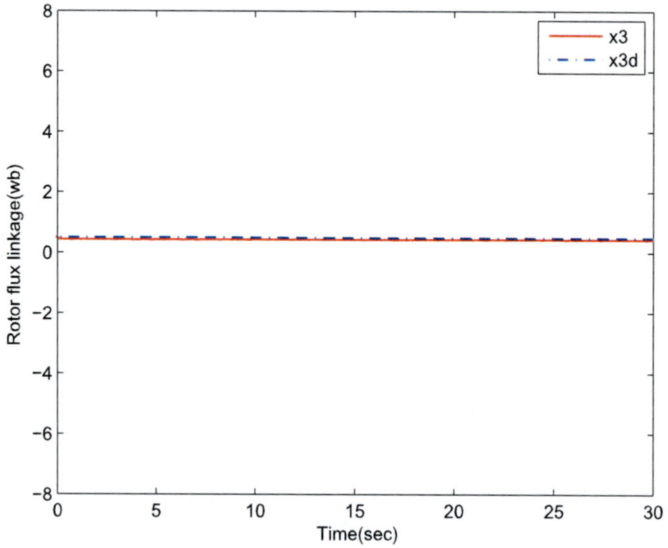

Fig. 5.8 Trajectories of the x_3 and x_{3d} for DSC

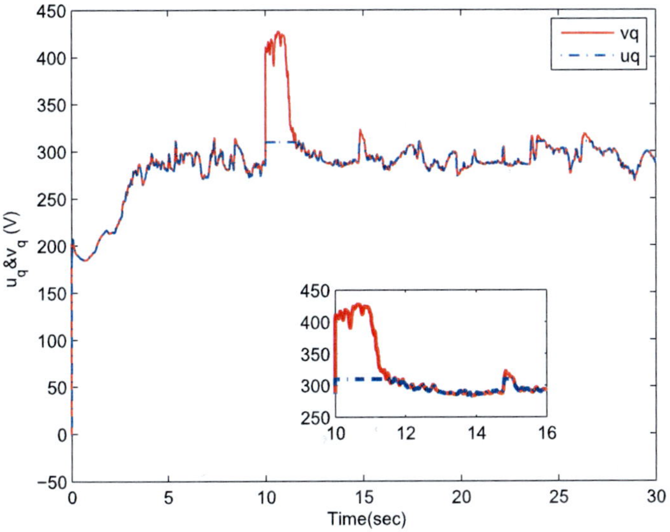

Fig. 5.9 Curves of the u_q and v_q for DSC

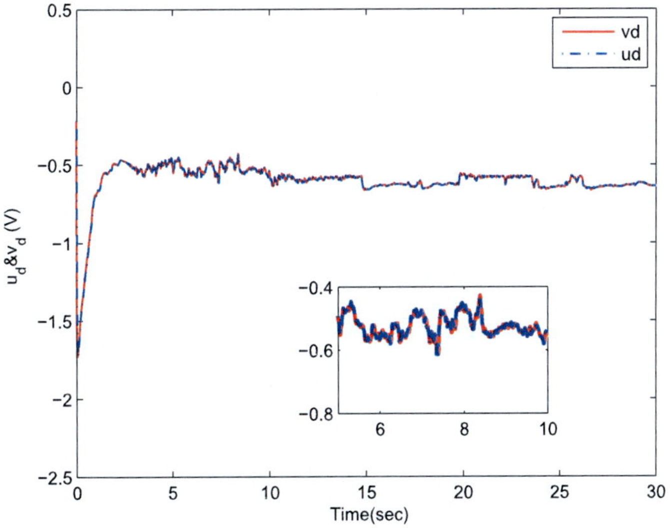

Fig. 5.10 Curves of the u_d and v_d for DSC

5.5 Conclusion

Based on command filtered technology, an adaptive fuzzy controller is constructed for IM stochastic nonlinear systems with stochastic disturbance and input saturation in this chapter. The FLS and CFC technologies are utilized to cope with unknown stochastic nonlinear functions and the "explosion of complexity" problem. The effectiveness and superiority of the proposed method are demonstrated by simulation results.

References

1. Chai, J.Y., Ho, Y.H., Chang, Y.C., Liaw, C.M.: On acoustic-noise-reduction control using random switching technique for switch-mode rectifiers in PMSM drive. IEEE Trans. Ind. Electron. **55**(3), 1295–1309 (2008)
2. Deng, H., Krstic, M.: Stabilization of stochastic nonlinear systems driven by noise of unknown covariance. IEEE Trans. Autom. Control **46**(8), 1237–1253 (2001)
3. Wang, H.Q., Chen, B., Liu, X.: Robust adaptive fuzzy tracking control for pure-feedback stochastic nonlinear systems with input constraints. IEEE Trans. Cybern. **43**(6), 2093–2104 (2013)
4. Yang, J., Wu, Z.J.: Stochastic position control for permanent magnet synchronous motor. In: Proceedings of Control and Decision Conference, Chinese, pp. 2192–2197 (2010)
5. Cui, G., Xu, S., Zhang, B., Lu, J., Li, Z., Zhang, Z.: Adaptive tracking control for uncertain switched stochastic nonlinear pure-feedback systems with unknown backlash-like hysteresis. J. Frankl. Inst. **354**(4), 1801–1818 (2017)
6. Alonge, F., Filippo, D., Sferlazza, A.: Sensorless control of induction-motor drive based on robust Kalman filter and adaptive speed estimation. IEEE Trans. Ind. Electron. **61**(3), 1444–1453 (2008)
7. Boulkroune, A., M'Saad, M., Farza, M.: Fuzzy approximation-based indirect adaptive controller for multi-input multi-output non-affine systems with unknown control direction. IET Control Theory Appl. **6**(17), 2619–2629 (2012)
8. Barambones, O., Alkorta, P.: Position control of the induction motor using an adaptive sliding-mode controller and observers. IEEE Trans. Ind. Electron. **61**(12), 6556–6565 (2014)
9. Azinheira, J.R., Moutinho, A.: Hover control of an UAV with backstepping design including input saturations. IEEE Trans. Control Syst. Technol. **16**(3), 517–526 (2008)
10. Liu, Y.J., Tong, S., Li, D.: Fuzzy adaptive control with state observer for a class of nonlinear discrete-time systems with input constraint. IEEE Trans. Fuzzy Syst. **24**(5), 1147–1158 (2016)
11. Liu, L.C., Ma, Y.M., Yu, J.P., Li, W., Wang, X.L.: Adaptive neural speed regulation control for induction motors stochastic nonlinear systems. ICIC Express Lett. **7**(8), 1747–1753 (2016)
12. Wang, H.Q., Liu, K., Liu, X., Chen, B., Lin, C.: Neural-based adaptive output-feedback control for a class of nonstrict-feedback stochastic nonlinear systems. IEEE Trans. Cybern. **45**(9), 1977–1987 (2015)
13. Tong, S., Li, Y., Li, Y., Liu, Y.: Observer-based adaptive fuzzy backstepping control for a class of stochastic nonlinear strict-feedback systems. IEEE Trans. Syst. Man Cybern. B Cybern. **41**(6), 1693–1704 (2011)
14. Wang, L., Mendel, J.M.: Fuzzy basis functions, universal approximation, and orthogonal least-squares learning. IEEE Trans. Neural Netw. **3**(5), 807–814 (1992)
15. Farrell, J., Polycarpou, M., Sharma, M., Dong, W.: Command filtered backstepping. IEEE Trans. Autom. Control **54**(6), 1391–1395 (2009)
16. Dong, W.J., Farrell, J., Polycarpou, M., Djapic, V., Sharma, M.: Command filtered adaptive backstepping. IEEE Trans. Control Syst. Technol. **20**(3), 566–580 (2012)

Chapter 6
Adaptive Fuzzy Dynamic Surface Control for IM with Iron Losses

A dynamic surface control (DSC) method combined with adaptive fuzzy backstepping technology [1] is developed for induction motor with iron losses in electric vehicle drive systems in this chapter. The DSC is utilized to overcome the "explosion of complexity" issue of classical backstepping. The fuzzy systems are used to approximate unknown nonlinear functions and the adaptive backstepping is employed to design controllers. The proposed control method can guarantee all the closed-loop signals are bounded. Simulation results illustrate its effectiveness.

6.1 Introduction

Recently, the electric vehicle has been an important branch of the automotive industry because of its great significance to energy security and environmental protection.

Induction motor (IM) has been increasingly applied in electric vehicles due to their remarkable advantages such as simple structure, high reliability, ruggedness and low cost. Many researchers have developed some nonlinear control techniques [2–13] to achieve high performance for IM drive systems. However, the above approaches have not considered the effect of iron losses for the IM drive systems. When the IM is used for electric vehicles working in light load condition for a long time, the system will generate too many iron losses which affect the whole system control property. When the electric vehicle is in a high speed, IM also causes a large amount of iron losses. It can't realize accurate control by using the above control methods because they all neglect iron losses.

In this chapter, an adaptive fuzzy DSC approach via backstepping is presented for position tracking control of the IM used for electric vehicle drive systems. Fuzzy systems are applied to cope with the nonlinearities and the DSC approach is used to solve the problem of "explosion of complexity".

J. Yu et al., *Intelligent Backstepping Control for the Alternating-Current Drive Systems*, Studies in Systems, Decision and Control 349,
https://doi.org/10.1007/978-3-030-67723-7_6

6.2 Mathematical Model and Preliminaries

Consider the dynamic mathematical model of the IM with iron losses in $d-q$ axis as follows [14, 15]:

$$
\begin{aligned}
\frac{d\Theta}{dt} &= \omega_r, \\
\frac{d\omega_r}{dt} &= \frac{n_p L_m}{L_{1r} J}\psi_d i_{qm} - \frac{T_L}{J}, \\
\frac{di_{qm}}{dt} &= \frac{R_{fe}}{L_m} i_{qs} - \frac{(L_m + L_{1r})R_{fe}}{L_{1r} L_m} i_{qm} + i_{dm}\omega_r + \frac{L_m R_r}{L_{1r}} \frac{i_{qm} i_{dm}}{\psi_d}, \\
\frac{di_{qs}}{dt} &= -\frac{R_s + R_{fe}}{L_{1s}} i_{qs} + \frac{L_m R_r}{L_{1r}} \frac{i_{ds} i_{qm}}{\psi_d} + i_{ds}\omega_r + \frac{(L_m + L_{1r})R_{fe}}{L_{1r} L_{1s}} i_{qm} + \frac{1}{L_{1s}} u_{qs}, \\
\frac{d\psi_d}{dt} &= -\frac{R_r}{L_{1r}}\psi_d + \frac{L_m}{L_{1r}} R_r i_{dm}, \\
\frac{di_{dm}}{dt} &= \frac{R_{fe}}{L_m} i_{ds} + \frac{R_{fe}}{L_{1r} L_m}\psi_d - \frac{(L_m + L_{1r})R_{fe}}{L_{1r} L_m} i_{dm} + \frac{L_m R_r}{L_{1r}} \frac{i_{qm}^2}{\psi_d} + i_{qm}\omega_r, \\
\frac{di_{ds}}{dt} &= -\frac{R_s + R_{fe}}{L_{1s}} i_{ds} + \frac{L_m R_r}{L_{1r}} \frac{i_{qs} i_{qm}}{\psi_d} + i_{qs}\omega_r + \frac{(L_m + L_{1r})R_{fe}}{L_{1r} L_{1s}} i_{dm}, \\
&\quad - \frac{R_{fe}}{L_{1s}^2}\psi_d + \frac{1}{L_{1s}} u_{ds},
\end{aligned}
$$

where Θ represents the rotor position, ω_r is the rotor angular velocity, T_L denotes the load torque, J and ψ_d stand for inertia and rotor flux linkage, respectively. n_p denotes pole pairs; i_{dm} and i_{qm} are $d-q$ axis exciting currents; i_{ds} and i_{qs} stand for the $d-q$ axis currents; R_r and R_s represent the resistance of rotor and stator; L_{1s} and L_{1r} stand for the inductance of stator and rotor; R_{fe} denotes the iron loss resistance; u_{ds} and u_{qs} mean the voltages of $d-q$ axis; L_m denotes mutual inductance. For calculation convenience, several notations can be constructed as:

$$
\begin{aligned}
& x_1 = \Theta, x_2 = \omega_r, x_3 = i_{qm}, x_4 = i_{qs}, x_5 = \varphi_d, x_6 = i_{dm}, x_7 = i_{ds}, \\
& a_1 = \frac{n_p L_m}{L_{1r}}, b_1 = \frac{R_{fe}}{L_m}, b_2 = \frac{(L_m + L_{1r})R_{fe}}{L_{1r} L_m}, b_3 = \frac{L_m R_r}{L_{1r}}, \\
& c_1 = \frac{1}{L_{1s}}, c_2 = \frac{R_s + R_{fe}}{L_{1s}}, c_3 = \frac{L_m R_r}{L_{1r}}, c_4 = \frac{(L_m + L_{1r})R_{fe}}{L_{1r} L_{1s}}, \\
& c_5 = \frac{R_{fe}}{L_{1s}^2}, d_1 = -\frac{R_r}{L_{1r}}, d_2 = \frac{L_m}{L_{1r}}, d_3 = \frac{R_{fe}}{L_m L_{1r}}.
\end{aligned}
$$

With the above notations, the original model can be transformed into

$$\begin{cases} \dot{x}_1 = x_2, \\ \dot{x}_2 = \frac{1}{J}a_1x_3x_5 - \frac{T_L}{J}, \\ \dot{x}_3 = b_1x_4 - b_2x_3 + b_3\frac{x_3x_6}{x_5} + x_2x_6, \\ \dot{x}_4 = c_1u_{qs} - c_2x_4 + x_2x_7 + c_3\frac{x_3x_7}{x_5} + c_4x_3, \\ \dot{x}_5 = d_1x_5 + d_2x_6, \\ \dot{x}_6 = b_1x_7 + d_3x_5 - b_2x_6 + b_3\frac{x_3^2}{x_5} + x_2x_3, \\ \dot{x}_7 = c_1u_{ds} - c_2x_7 + c_3\frac{x_3x_4}{x_5} + x_2x_4 - c_5x_5 + c_4x_6. \end{cases} \tag{6.1}$$

Lemma 6.1 *[16] Let $f(x)$ be a continuous function defined on a compact set Ω. Then for any scalar $\varepsilon > 0$, there exists a fuzzy logic system $W^T S(x)$ such that*

$$\sup_{x\in\Omega} \left| f(x) - W^T S(x) \right| \leq \varepsilon,$$

where $W = [W_1, ..., W_N]^T$ is the ideal constant weight vector, and $S(x) = [p_1(x), p_2(x), ..., p_N(x)]^T / \sum_{i=1}^N p_i(x)$ is the basis function vector, with $N > 1$ being the number of the fuzzy rules and p_i are chosen as Gaussian functions, i.e., for $i = 1, 2, ..., N$, $p_i(x) = \exp[\frac{-(x-\mu_i)^T(x-\mu_i)}{\eta_i^2}]$ where $\mu_i = [\mu_{i1}, \mu_{i2}, ..., \mu_{in}]^T$ is the center vector, and η_i is the width of the Gaussian function.

6.3 Adaptive Fuzzy DSC Design with Backstepping

This section is devoted to provide the DSC approach with fuzzy approximation to construct controllers for IM with iron losses described by (6.1).

Step 1: Define the tracking error variable $z_1 = x_1 - x_{1d}$ with x_{1d} being the desired signal. For the first subsystem of (6.1), choose the Lyapunov function candidate as $V_1 = \frac{1}{2}z_1^2$. Then,

$$\dot{V}_1 = z_1\dot{z}_1 = z_1(x_2 - \dot{x}_{1d}). \tag{6.2}$$

Design the virtual control as

$$\alpha_1 = -k_1z_1 + \dot{x}_{1d}, \tag{6.3}$$

with $k_1 > 0$ being the control gain. At present stage, we introduce a new variable α_{1d} and a time constant ϵ_1, and then let α_1 pass through a first-order filter to obtain α_{1d} as

$$\epsilon_1\dot{\alpha}_{1d} + \alpha_{1d} = \alpha_1, \alpha_{1d}(0) = \alpha_1(0). \tag{6.4}$$

Construct $z_2 = x_2 - \alpha_{1d}$. From (6.3) and (6.4), we can get

$$\begin{aligned} \dot{V}_1 &= z_1z_2 + z_1(\alpha_{1d} - \alpha_1 + \alpha_1) - z_1\dot{x}_{1d} \\ &= -k_1z_1^2 + z_1z_2 + z_1(\alpha_{1d} - \alpha_1). \end{aligned} \tag{6.5}$$

Step 2: From the first step, the differential form of z_2 can be written as:

$$\dot{z}_2 = \dot{x}_2 - \dot{\alpha}_{1d} = \frac{1}{J}a_1x_3x_5 - \frac{T_L}{J} - \dot{\alpha}_{1d}. \tag{6.6}$$

Similarly, choose $V_2 = V_1 + \frac{J}{2}z_2^2$. Differentiating V_2 along with (6.5) and (6.3) gives

$$\dot{V}_2 = -k_1z_1^2 + z_1(\alpha_{1d} - \alpha_1) + z_2(z_1 + a_1x_3x_5 - T_L - J\dot{\alpha}_{1d}). \tag{6.7}$$

Remark 6.2 In this chapter, the load torque T_L is considered to be unknown in IM drive system and its upper limit is assumed to be $d > 0$, that is, $|T_L| \leq d$. ♦

By using Young's inequality, we can gain $-z_2T_L \leq \frac{1}{2\varepsilon_1^2}z_2^2 + \frac{1}{2}\varepsilon_1^2d^2$ with ε_1 being an arbitrary small positive constant. Then we have

$$\dot{V}_2 \leq -k_1z_1^2 + \frac{1}{2}\varepsilon_1^2d^2 + z_1(\alpha_{1d} - \alpha_1) + z_2(z_1 + a_1x_3x_5 + \frac{1}{2\varepsilon_1^2}z_2 - J\dot{\alpha}_{1d}). \tag{6.8}$$

For simplifying the analysis process, (6.8) can be rewritten as

$$\dot{V}_2 \leq -k_1z_1^2 + \frac{1}{2}\varepsilon_1^2d^2 + z_1(\alpha_{1d} - \alpha_1) + z_2(x_3 + z_1 + a_1x_3x_5 + \frac{1}{2\varepsilon_1^2}z_2 - J\dot{\alpha}_{1d} - x_3). \tag{6.9}$$

The following fuzzy logic system is introduced as:

$$f_2(Z) = z_1 + a_1x_3x_5 + \frac{1}{2\varepsilon_1^2}z_2 - x_3 = W_2^TS_2(Z) + \delta_2(Z),$$

$$Z = [x_1, x_2, x_3, x_4, x_5, x_6, x_7, x_{1d}, \dot{x}_{1d}]^T,$$

with $\delta_2(Z)$ being the approximation error and satisfies $|\delta_2| \leq \varepsilon_2$. Nevertheless, ε_2 is considered to be an unknown positive constant. By using a straightforward calculation, we can get the following inequality

$$z_2f_2 \leq \frac{1}{2l_2^2}z_2^2||W_2||^2S_2^TS_2 + \frac{1}{2}l_2^2 + \frac{1}{2}z_2^2 + \frac{1}{2}\varepsilon_2^2. \tag{6.10}$$

At this step, we construct the virtual control function as

$$\alpha_2 = -k_2z_2 - \frac{1}{2}z_2 - \frac{1}{2l_2^2}z_2\hat{\theta}S_2^TS_2 + \hat{J}\dot{\alpha}_{1d}, \tag{6.11}$$

with $k_2 > 0$. Define a new state variable α_{2d} and a time constant ϵ_2. Then let α_2 pass through a first-order filter to obtain α_{2d} as

$$\epsilon_2\dot{\alpha}_{2d} + \alpha_{2d} = \alpha_2, \alpha_{2d}(0) = \alpha_2(0). \tag{6.12}$$

Introducing $z_3 = x_3 - \alpha_{2d}$ and substituting (6.10), (6.11) and (6.12) into (6.9), one has

$$\dot{V}_2 \leq -k_1 z_1^2 - k_2 z_2^2 + \frac{1}{2}\varepsilon_1^2 d^2 + z_1(\alpha_{1d} - \alpha_1) + z_2(\alpha_{2d} - \alpha_2) + \frac{1}{2}l_2^2$$
$$+\frac{1}{2}\varepsilon_2^2 + \frac{1}{2l_2^2}z_2^2(||W_2||^2 - \hat{\theta})S_2^T S_2 + z_2 z_3 + z_2(\hat{J} - J)\dot{\alpha}_{1d}. \quad (6.13)$$

Step 3: The time derivative of z_3 is

$$\dot{z}_3 = \dot{x}_3 - \dot{\alpha}_{2d} = b_1 x_4 - b_2 x_3 + b_3 \frac{x_3 x_6}{x_5} + x_2 x_6 - \dot{\alpha}_{2d}. \quad (6.14)$$

The Lyapunov function candidate V_3 is defined as $V_3 = V_2 + \frac{1}{2}z_3^2$, and the differential form of V_3 along with (6.13) and (6.14) is

$$\dot{V}_3 \leq -k_1 z_1^2 - k_2 z_2^2 + \frac{1}{2}\varepsilon_1^2 d^2 + z_1(\alpha_{1d} - \alpha_1) + z_2(\alpha_{2d} - \alpha_2) + \frac{1}{2}l_2^2$$
$$+\frac{1}{2}\varepsilon_2^2 + \frac{1}{2l_2^2}z_2^2(||W_2||^2 - \hat{\theta})S_2^T S_2 + z_3(z_2 + b_1 x_4 - b_2 x_3$$
$$+b_3 \frac{x_3 x_6}{x_5} + x_2 x_6 - \dot{\alpha}_{2d}) + z_2(\hat{J} - J)\dot{\alpha}_{1d}, \quad (6.15)$$

where $f_3(Z) = z_2 - b_2 x_3 + b_3 \frac{x_3 x_6}{x_5} + x_2 x_6 - \dot{\alpha}_{2d} = W_3^T S_3(Z) + \delta_3(Z)$ with $|\delta_3| \leq \varepsilon_3$.

Similarly, by Lemma 6.1, for given $\varepsilon_3 > 0$, we can get the following inequality

$$z_3 f_3 \leq \frac{1}{2l_3^2} z_3^2 \, \|W_3\|^2 \, S_3^T S_3 + \frac{1}{2}l_3^2 + \frac{1}{2}z_3^2 + \frac{1}{2}\varepsilon_3^2. \quad (6.16)$$

Construct the virtual control law of this step as

$$\alpha_3 = \frac{1}{b_1}(-k_3 z_3 - \frac{1}{2}z_3 - \frac{1}{2l_3^2}z_3 \hat{\theta} S_3^T S_3), \quad (6.17)$$

with $k_3 > 0$. Similarly, a new state variable α_{3d} and a time constant ϵ_3 are introduced to pass through a filter to obtain α_{3d} as

$$\epsilon_3 \dot{\alpha}_{3d} + \alpha_{3d} = \alpha_3, \, \alpha_{3d}(0) = \alpha_3(0). \quad (6.18)$$

Let $z_4 = x_4 - \alpha_{3d}$ and replacing (6.16), (6.17) and (6.18) into (6.15) results in

$$\begin{aligned}\dot{V}_3 \leq & -k_1 z_1^2 - k_2 z_2^2 - k_3 z_3^2 + \frac{1}{2}\varepsilon_1^2 d^2 + z_1(\alpha_{1d} - \alpha_1) + z_2(\alpha_{2d} - \alpha_2) \\ & + b_1 z_3(\alpha_{3d} - \alpha_3) + \frac{1}{2}l_2^2 + \frac{1}{2}\varepsilon_2^2 + \frac{1}{2}l_3^2 + \frac{1}{2}\varepsilon_3^2 \\ & + \frac{1}{2l_2^2} z_2^2(||W_2||^2 - \hat{\theta}) S_2^T S_2 + \frac{1}{2l_3^2} z_3^2(||W_3||^2 - \hat{\theta}) S_3^T S_3 \\ & + b_1 z_3 z_4 + z_2(\hat{J} - J)\dot{\alpha}_{1d}. \end{aligned} \tag{6.19}$$

Step 4: The time derivative of z_4 can be calculated as follows:

$$\dot{z}_4 = \dot{x}_4 - \dot{\alpha}_{3d} = c_1 u_{qs} - c_2 x_4 + x_2 x_7 + c_3 \frac{x_3 x_7}{x_5} + c_4 x_3 - \dot{\alpha}_{3d}.$$

Choose $V_4 = V_3 + \frac{1}{2} z_4^2$, then

$$\begin{aligned}\dot{V}_4 \leq & -k_1 z_1^2 - k_2 z_2^2 - k_3 z_3^2 + \frac{1}{2}\varepsilon_1^2 d^2 + z_1(\alpha_{1d} - \alpha_1) + z_2(\alpha_{2d} - \alpha_2) \\ & + b_1 z_3(\alpha_{3d} - \alpha_3) + \frac{1}{2}l_2^2 + \frac{1}{2}\varepsilon_2^2 + \frac{1}{2}l_3^2 + \frac{1}{2}\varepsilon_3^2 \\ & + \frac{1}{2l_2^2} z_2^2(||W_2||^2 - \hat{\theta}) S_2^T S_2 + \frac{1}{2l_3^2} z_3^2(||W_3||^2 - \hat{\theta}) S_3^T S_3 \\ & + z_2(\hat{J} - J)\dot{\alpha}_{1d} + z_4(b_1 z_3 + c_1 u_{qs} - c_2 x_4 + x_2 x_7 \\ & + c_3 \frac{x_3 x_7}{x_5} + c_4 x_3 - \dot{\alpha}_{3d}), \end{aligned} \tag{6.20}$$

where $f_4(Z) = b_1 z_3 - c_2 x_4 + x_2 x_7 + c_3 \frac{x_3 x_7}{x_5} + c_4 x_3 - \dot{\alpha}_{3d} = W_4^T S_4(Z) + \delta_4(Z)$ and $|\delta_4| \leq \varepsilon_4$.

Remark 6.3 Notice that f_4 contains the nonlinear terms $x_2 x_7$ and $c_3 \frac{x_3 x_7}{x_5}$ as well as the time derivative of α_{3d}. However, these issues are hard to be dealt with in classical backstepping and the structure of control law u_{qs} will be very complex. To overcome these problems, we will introduce fuzzy logic systems to approximate nonlinear functions to simplify the design process of the controllers. As shown later, the controller u_{qs} will have a simpler structure than the traditional one's. ♦

Similarly, by using Lemma 6.1, for any given $\varepsilon_4 > 0$, we can obtain the following inequality

$$z_4 f_4 \leq \frac{1}{2l_4^2} z_4^2 ||W_4||^2 S_4^T S_4 + \frac{1}{2}l_4^2 + \frac{1}{2}z_4^2 + \frac{1}{2}\varepsilon_4^2. \tag{6.21}$$

For the control design of this system, we construct the control law u_{qs} as

$$u_{qs} = \frac{1}{c_1}(-k_4 z_4 - \frac{1}{2} z_4 - \frac{1}{2l_4^2} z_4 \hat{\theta} S_4^T S_4), \tag{6.22}$$

with $k_4 > 0$. By the Eqs. (6.21) and (6.22), we can get

$$\begin{aligned}\dot{V}_4 \leq & -\sum_{i=1}^{4} k_i z_i^2 + \frac{1}{2}\varepsilon_1^2 d^2 + z_1(\alpha_{1d} - \alpha_1) + z_2(\alpha_{2d} - \alpha_2) + b_1 z_3(\alpha_{3d} - \alpha_3) \\ & + \frac{1}{2}l_2^2 + \frac{1}{2}\varepsilon_2^2 + \frac{1}{2}l_3^2 + \frac{1}{2}\varepsilon_3^2 + \frac{1}{2}l_4^2 + \frac{1}{2}\varepsilon_4^2 + \frac{1}{2l_2^2}z_2^2(||W_2||^2 - \hat{\theta})S_2^T S_2 \\ & + \frac{1}{2l_3^2}z_3^2(||W_3||^2 - \hat{\theta})S_3^T S_3 + \frac{1}{2l_4^2}z_4^2(||W_4||^2 - \hat{\theta})S_4^T S_4 \\ & + z_2(\hat{J} - J)\dot{\alpha}_{1d}. \end{aligned} \tag{6.23}$$

Step 5: Define the second reference signal as x_{5d}, then this tracking error variable can be given as $z_5 = x_5 - x_{5d}$. Lyapunov candidate function V_5 is defined by $V_5 = V_4 + \frac{1}{2}z_5^2$, then we can get

$$\dot{V}_5 \leq \dot{V}_4 + z_5(d_1 x_5 + d_2 x_6 - \dot{x}_{5d}). \tag{6.24}$$

Similarly, construct the virtual control law of this step as follows:

$$\alpha_4 = (-k_5 z_5 + \dot{x}_{5d} - d_1 x_5)/d_2, \tag{6.25}$$

with $k_5 > 0$. Define a new state variable α_{4d} and a time constant ϵ_4. Then letting α_4 pass through a first-order filter, we can obtain α_{4d} as

$$\epsilon_4 \dot{\alpha}_{4d} + \alpha_{4d} = \alpha_4, \; \alpha_{4d}(0) = \alpha_4(0). \tag{6.26}$$

In addition, let $z_6 = x_6 - \alpha_{4d}$ and replacing (6.25) and (6.26) into (6.24) results in

$$\dot{V}_5 = \dot{V}_4 - k_5 z_5^2 + d_2 z_5(\alpha_{4d} - \alpha_4) + d_2 z_5 z_6. \tag{6.27}$$

Step 6: The differential form of z_6 is

$$\dot{z}_6 = \dot{x}_6 - \dot{\alpha}_{4d} = b_1 x_7 + d_3 x_5 - b_2 x_6 + b_3 \frac{x_3^2}{x_5} + x_2 x_3 - \dot{\alpha}_{4d}. \tag{6.28}$$

Choose $V_6 = V_5 + \frac{1}{2}z_6^2$. Then computing its time derivative form along with (6.27) and (6.28), we can obtain

$$\begin{aligned}\dot{V}_6 \leq & \dot{V}_4 - k_5 z_5^2 + d_2 z_5(\alpha_{4d} - \alpha_4) + d_2 z_5 z_6 + z_6(b_1 x_7 \\ & + d_3 x_5 - b_2 x_6 + b_3 \frac{x_3^2}{x_5} + x_2 x_3 - \dot{\alpha}_{4d}), \end{aligned} \tag{6.29}$$

where $f_6(Z) = d_2z_5 + d_3x_5 - b_2x_6 + b_3\frac{x_3^2}{x_5} + x_2x_3 - \dot{\alpha}_{4d} = W_6^T S_6(Z) + \delta_6(Z)$ with $|\delta_6| \leq \varepsilon_6$.

By Lemma 6.1, for given $\varepsilon_6 > 0$, we can obtain the following inequality

$$z_6 f_6 \leq \frac{1}{2l_6^2} z_6^2 ||W_6||^2 S_6^T S_6 + \frac{1}{2} l_6^2 + \frac{1}{2} z_6^2 + \frac{1}{2} \varepsilon_6^2. \tag{6.30}$$

Construct the virtual control law of this step as:

$$\alpha_5 = \frac{1}{b_1}(-k_6 z_6 - \frac{1}{2} z_6 - \frac{1}{2l_6^2} z_6 \hat{\theta} S_6^T S_6), \tag{6.31}$$

with $k_6 > 0$. Introduce a new state variable α_{5d} and a time constant ϵ_5. Then let α_5 pass through a first-order filter, we can gain α_{5d} as

$$\epsilon_5 \dot{\alpha}_{5d} + \alpha_{5d} = \alpha_5, \alpha_{5d}(0) = \alpha_5(0). \tag{6.32}$$

Substituting (6.30), (6.31) and (6.32) into (6.29) and defining $z_7 = x_7 - \alpha_{5d}$ result in

$$\begin{aligned} \dot{V}_6 \leq\; & \dot{V}_4 - k_5 z_5^2 - k_6 z_6^2 + d_2 z_5(\alpha_{4d} - \alpha_4) + b_1 z_6(\alpha_{5d} - \alpha_5) + b_1 z_6 z_7 \\ & + \frac{1}{2l_6^2} z_6^2 (\| W_6^T \|^2 - \hat{\theta}) S_6^T S_6 + \frac{1}{2} l_6^2 + \frac{1}{2} \varepsilon_6^2. \end{aligned} \tag{6.33}$$

Step 7: During this step, another real control law u_{ds} is designed. Differentiating z_7 results in the following equation

$$\dot{z}_7 = \dot{x}_7 - \dot{\alpha}_{5d} = c_1 u_{ds} - c_2 x_7 + c_3 \frac{x_3 x_4}{x_5} + x_2 x_4 - c_5 x_5 + c_4 x_6 - \dot{\alpha}_{5d}.$$

Now choose $V_7 = V_6 + \frac{1}{2} z_7^2$. Then the time derivative form can be written as

$$\begin{aligned} \dot{V}_7 \leq\; & \dot{V}_4 - k_5 z_5^2 - k_6 z_6^2 + d_2 z_5(\alpha_{4d} - \alpha_4) + b_1 z_6(\alpha_{5d} - \alpha_5) + b_1 z_6 z_7 \\ & + \frac{1}{2l_6^2} z_6^2 (\| W_6^T \|^2 - \hat{\theta}) S^2 + \frac{1}{2} l_6^2 + \frac{1}{2} \varepsilon_6^2 + z_7(c_1 u_{ds} - c_2 x_7 \\ & + c_3 \frac{x_3 x_4}{x_5} + x_2 x_4 - c_5 x_5 + c_4 x_6 - \dot{\alpha}_{5d}), \end{aligned} \tag{6.34}$$

where $f_7(Z) = b_1 z_6 - c_2 x_7 + c_3 \frac{x_3 x_4}{x_5} + x_2 x_4 - c_5 x_5 + c_4 x_6 - \dot{\alpha}_{5d} = W_7^T S_7(Z) + \delta_7(Z)$ with $|\delta_7| \leq \varepsilon_7$.

Similarly, by Lemma 6.1, for given $\varepsilon_7 > 0$, we can obtain

$$z_7 f_7 \leq \frac{1}{2l_7^2} z_7^2 ||W_7||^2 S_7^T S_7 + \frac{1}{2} l_7^2 + \frac{1}{2} z_7^2 + \frac{1}{2} \varepsilon_7^2. \tag{6.35}$$

Construct the real control law u_{ds} as

$$u_{ds} = \frac{1}{c_1}(-k_7 z_7 - \frac{1}{2} z_7 - \frac{1}{2l_7^2} z_7 \hat{\theta} S_7^T S_7), \tag{6.36}$$

with $k_7 > 0$.

Define $\theta = \max\{||W_2||^2, ||W_3||^2, ||W_4||^2, ||W_6||^2, ||W_7||^2\}, \tilde{\theta} = \hat{\theta} - \theta, \tilde{J} = \hat{J} - J$. Then by the Eqs. (6.35) and (6.36), we can get

$$\begin{aligned}
\dot{V}_7 \le & -\sum_{i=1}^{7} k_i z_i^2 + \frac{1}{2}\varepsilon_1^2 d^2 + z_1(\alpha_{1d} - \alpha_1) + z_2(\alpha_{2d} - \alpha_2) + b_1 z_3(\alpha_{3d} - \alpha_3) \\
& + d_2 z_5(\alpha_{4d} - \alpha_4) + b_1 z_6(\alpha_{5d} - \alpha_5) + \frac{1}{2} l_2^2 + \frac{1}{2}\varepsilon_2^2 + \frac{1}{2} l_3^2 \\
& + \frac{1}{2}\varepsilon_3^2 + \frac{1}{2} l_4^2 + \frac{1}{2}\varepsilon_4^2 + \frac{1}{2} l_6^2 + \frac{1}{2}\varepsilon_6^2 + \frac{1}{2} l_7^2 + \frac{1}{2}\varepsilon_7^2 \\
& - \frac{1}{2l_2^2} z_2^2 \tilde{\theta} S_2^T S_2 - \frac{1}{2l_3^2} z_3^2 \tilde{\theta} S_3^T S_3 - \frac{1}{2l_4^2} z_4^2 \tilde{\theta} S_4^T S_4 \\
& - \frac{1}{2l_6^2} z_6^2 \tilde{\theta} S_6^T S_6 - \frac{1}{2l_7^2} z_7^2 \tilde{\theta} S_7^T S_7 + z_2 \tilde{J} \dot{\alpha}_{1d}.
\end{aligned} \tag{6.37}$$

Define $y_i = \alpha_{id} - \alpha_i, i = 1, \ldots, 5$. The following equations can be obtained:

$$\dot{y}_i = \dot{\alpha}_{id} - \dot{\alpha}_i = -\frac{\alpha_{id} - \alpha_i}{\epsilon_i} - \dot{\alpha}_i = -\frac{y_i}{\epsilon_i} + B_i, \tag{6.38}$$

with $B_i = \dot{\alpha}_i$. Choose the following Lyapunov function: $V = V_7 + \frac{1}{2}y_1^2 + \frac{1}{2}y_2^2 + \frac{1}{2}y_3^2 + \frac{1}{2}y_4^2 + \frac{1}{2}y_5^2 + \frac{\tilde{\theta}^2}{2r_1} + \frac{\tilde{J}^2}{2r_2}$, with r_1 and r_2 being positive constants.

Compute the time derivative of V, then according to (6.37) and (6.38), it can be rewritten as:

$$\begin{aligned}
\dot{V} \le & -\sum_{i=1}^{7} k_i z_i^2 + \frac{1}{2}\varepsilon_1^2 d^2 + z_1 y_1 + z_2 y_2 + b_1 z_3 y_3 + d_2 z_5 y_4 + b_1 z_6 y_5 \\
& + \frac{1}{2} l_2^2 + \frac{1}{2}\varepsilon_2^2 + \frac{1}{2} l_3^2 + \frac{1}{2}\varepsilon_3^2 + \frac{1}{2} l_4^2 + \frac{1}{2}\varepsilon_4^2 + \frac{1}{2} l_6^2 + \frac{1}{2}\varepsilon_6^2 + \frac{1}{2} l_7^2 \\
& + \frac{1}{2}\varepsilon_7^2 + \sum_{i=1}^{5} y_i \dot{y}_i + \frac{\tilde{\theta}}{r_1}(\dot{\hat{\theta}} - \frac{r_1}{2l_2^2} z_2^2 S_2^T S_2 - \frac{r_1}{2l_3^2} z_3^2 S_3^T S_3 \\
& - \frac{r_1}{2l_4^2} z_4^2 S_4^T S_4 - \frac{r_1}{2l_6^2} z_6^2 S_6^T S_6 - \frac{r_1}{2l_7^2} z_7^2 S_7^T S_7) \\
& + \frac{\tilde{J}}{2r_2}(\dot{\hat{J}} + r_2 z_2 \dot{\alpha}_{1d}).
\end{aligned} \tag{6.39}$$

By use of (6.39), we construct the adaptive laws as:

$$
\begin{aligned}
\dot{\hat{\theta}} &= \frac{r_1}{2l_2^2} z_2^2 S_2^T S_2 + \frac{r_1}{2l_3^2} z_3^2 S_3^T S_3 + \frac{r_1}{2l_4^2} z_4^2 S_4^T S_4 \\
&\quad + \frac{r_1}{2l_6^2} z_6^2 S_6^T S_6 + \frac{r_1}{2l_7^2} z_7^2 S_7^T S_7 - m_1 \hat{\theta}, \\
\dot{\hat{J}} &= -r_2 z_2 \dot{\alpha}_{1d} - m_2 \hat{J},
\end{aligned} \tag{6.40}
$$

with m_1, m_2 and $l_i (i = 2, 3, 4, 6, 7)$ being positive constants.

Remark 6.4 It is worth pointing out that one of the traditional backstepping problems called "explosion of complexity" is overcome by introducing the DSC technology. Moreover, fuzzy systems can cope with the unknown system parameters and make the controllers structure much simpler which makes the proposed control scheme more appropriate to real world applications. ♦

Remark 6.5 From (6.22) and (6.36), we can obtain that the real controllers u_{ds} and u_{qs} have simpler structure which indicates that the proposed fuzzy adaptive DSC control scheme will be more suitable for industrial applications. A simulation comparison between the DSC controllers and the traditional ones is given in Sect. 6.4 to illustrate this point. ♦

Theorem 6.6 *Consider system (6.1) and the given reference signals x_{1d} and x_{5d}, the DSC based adaptive fuzzy controllers (6.22), (6.36), the virtual controllers (6.3), (6.11), (6.17), (6.25), (6.31) and the adaptive laws (6.40) can guarantee that the convergence of the tracking errors falls into a sufficiently small neighborhood of origin and all the closed-loop variables are bounded.*

Proof To confirm the stability of the proposed control scheme, substituting (6.40) into (6.39), one has

$$
\begin{aligned}
\dot{V} \le & -\sum_{i=1}^{7} k_i z_i^2 + \frac{1}{2}\varepsilon_1^2 d^2 + z_1 y_1 + z_2 y_2 + b_1 z_3 y_3 + d_2 z_5 y_4 + b_1 z_6 y_5 \\
& + \frac{1}{2} l_2^2 + \frac{1}{2}\varepsilon_2^2 + \frac{1}{2} l_3^2 + \frac{1}{2}\varepsilon_3^2 + \frac{1}{2} l_4^2 + \frac{1}{2}\varepsilon_4^2 + \frac{1}{2} l_6^2 + \frac{1}{2}\varepsilon_6^2 \\
& + \frac{1}{2} l_7^2 + \frac{1}{2}\varepsilon_7^2 + \sum_{i=1}^{5} y_i \dot{y}_i - \frac{m_1 \tilde{\theta}\hat{\theta}}{r_1} - \frac{m_2 \tilde{J}\hat{J}}{r_2}.
\end{aligned} \tag{6.41}
$$

Known from dynamic surface technology, $|B_i|$ has a maximum B_{iM} on compact set $|\Omega_i|, i = 1, 2, 3, 4, 5, |B_i| \le B_{iM}$. Hence, $y_i \dot{y}_i \le -\frac{y_i^2}{\epsilon_i} + |B_{iM}||y_i| \le -\frac{y_i^2}{\epsilon_i} + \frac{1}{2\tau} B_{iM}^2 y_i^2 + \frac{\tau}{2}$ with $\tau > 0$. With Young's inequality, we can get the following formula:

$$z_1 y_1 \le \frac{1}{4} y_1^2 + z_1^2, z_2 y_2 \le \frac{1}{4} y_2^2 + z_2^2, b_1 z_3 y_3 \le \frac{b_1^2}{4} y_3^2 + z_3^2,$$
$$d_2 z_5 y_4 \le \frac{d_2^2}{4} y_4^2 + z_5^2, b_1 z_6 y_5 \le \frac{b_1^2}{4} y_5^2 + z_6^2, -\tilde{\theta}\hat{\theta} \le -\frac{\tilde{\theta}^2}{2} + \frac{\theta^2}{2},$$
$$-\tilde{J}\hat{J} \le -\frac{\tilde{J}^2}{2} + \frac{J^2}{2}.$$

Then, (6.41) can be rewritten as

$$\begin{aligned}\dot{V} \le& -(k_1-1)z_1^2 - (k_2-1)z_2^2 - (k_3-1)z_3^2 - k_4 z_4^2 - (k_5-1)z_5^2 \\ &-(k_6-1)z_6^2 - k_7 z_7^2 - \frac{m_1\tilde{\theta}^2}{2r_1} - \frac{m_2\tilde{J}^2}{2r_2} - (\frac{1}{\epsilon_1} - (\frac{1}{4} + \frac{1}{2\tau}B_{1M}^2))y_1^2 \\ &-(\frac{1}{\epsilon_2} - (\frac{1}{4} + \frac{1}{2\tau}B_{2M}^2))y_2^2 - (\frac{1}{\epsilon_3} - (\frac{b_1^2}{4} + \frac{1}{2\tau}B_{3M}^2))y_3^2 \\ &-(\frac{1}{\epsilon_4} - (\frac{d_2^2}{4} + \frac{1}{2\tau}B_{4M}^2))y_4^2 - (\frac{1}{\epsilon_5} - (\frac{b_1^2}{4} + \frac{1}{2\tau}B_{5M}^2))y_5^2 \\ &+\frac{1}{2}l_2^2 + \frac{1}{2}\varepsilon_2^2 + \frac{1}{2}l_3^2 + \frac{1}{2}\varepsilon_3^2 + \frac{1}{2}l_4^2 + \frac{1}{2}\varepsilon_4^2 + \frac{1}{2}l_6^2 + \frac{1}{2}\varepsilon_6^2 + \frac{1}{2}l_7^2 \\ &+\frac{1}{2}\varepsilon_7^2 + \frac{m_1\theta^2}{2r_1} + \frac{m_2 J^2}{2r_2} + \frac{5}{2}\tau + \frac{1}{2}\varepsilon_1^2 d^2 \le -a_0 V + b_0,\end{aligned} \tag{6.42}$$

where $a_0 = \min\left\{\begin{array}{c} 2(k_1-1), 2(k_2-1), 2(k_3-1), 2k_4, 2(k_5-1), 2(k_6-1), 2k_7, \\ m_1, m_2, 2(\frac{1}{\epsilon_1} - (\frac{1}{4} + \frac{1}{2\tau}B_{1M}^2)), 2(\frac{1}{\epsilon_2} - (\frac{1}{4} + \frac{1}{2\tau}B_{2M}^2)), \\ 2(\frac{1}{\epsilon_3} - (\frac{b_1^2}{4} + \frac{1}{2\tau}B_{3M}^2)), 2(\frac{1}{\epsilon_4} - (\frac{d_2^2}{4} + \frac{1}{2\tau}B_{4M}^2)), \\ 2(\frac{1}{\epsilon_5} - (\frac{b_1^2}{4} + \frac{1}{2\tau}B_{5M}^2)) \end{array}\right\}$
and $b_0 = \frac{1}{2}l_2^2 + \frac{1}{2}\varepsilon_2^2 + \frac{1}{2}l_3^2 + \frac{1}{2}\varepsilon_3^2 + \frac{1}{2}l_4^2 + \frac{1}{2}\varepsilon_4^2 + \frac{1}{2}l_6^2 + \frac{1}{2}\varepsilon_6^2 + \frac{1}{2}l_7^2 + \frac{1}{2}\varepsilon_7^2 + \frac{m_1\theta^2}{2r_1} + \frac{m_2 J^2}{2r_2} + \frac{5}{2}\tau + \frac{1}{2}\varepsilon_1^2 d^2$. *Equation* (6.42) indicates that

$$V(t) \le (V(t_0) - \frac{b_0}{a_0})e^{-a_0(t-t_0)} + \frac{b_0}{a_0} \le V(t_0) + \frac{b_0}{a_0}, \forall t \ge t_0. \tag{6.43}$$

As a result, $z_i (i = 1, 2, 3, 4, 5, 6, 7)$, $\tilde{J}$ and $\tilde{\theta}$ are in the compact set

$$\Omega = \left\{(z_i, \tilde{J}, \tilde{\theta}) | V \le V(t_0) + \frac{b_0}{a_0}, \forall t \ge t_0\right\}.$$

That is, every variable in this closed-loop system are bounded. Specifically, by using (6.43), one has

$$\lim_{t\to\infty} z_1^2 \le \frac{2b_0}{a_0}.$$

Remark 6.7 From the expressions of a_0 and b_0, the tracking error can be very small by choosing sufficiently large r_i and small enough ε_i and l_i after the parameters τ, ϵ_i, k_i and m_i are defined. The control precision can be improved by selecting the large enough k_i, but the computational load will increase as well. Moreover, decreasing ϵ_i decreases the tracking error, but for a first-order DSC filter, decreasing ϵ_i also increases the magnitude of the dynamic surface derivatives. ♦

■

6.4 A Comparison with the Traditional Adaptive Fuzzy Backstepping Design

This part will give the traditional adaptive fuzzy backstepping technique design according to [11].

Step 1: Consider the desired signal x_{1d}, then $z_1 = x_1 - x_{1d}$. By system (6.4), the time derivative form of this error can be written as: $\dot{z}_1 = x_2 - \dot{x}_{1d}$. Choose the Lyapunov function as $V_1 = \frac{1}{2}z_1^2$, then its differential is given as

$$\dot{V}_1 = z_1\dot{z}_1 = z_1(x_2 - \dot{x}_{1d}). \tag{6.44}$$

Define the virtual controller of the first step as

$$\alpha_1 = -k_1 z_1 + \dot{x}_{1d}, \tag{6.45}$$

where $k_1 > 0$ is the designed control parameter and $z_2 = x_2 - \alpha_1$. With (6.45), (6.44) can be replaced into the following equation.

$$\dot{V}_1 = -k_1 z_1^2 + z_1 z_2. \tag{6.46}$$

Step 2: The time derivative of z_2 is computed by

$$\dot{z}_2 = \dot{x}_2 - \dot{\alpha}_1 = \frac{1}{J}a_1 x_3 x_5 - \frac{T_L}{J} - \dot{\alpha}_1. \tag{6.47}$$

Choose $V_2 = V_1 + \frac{J}{2}z_2^2$ and differentiate V_2 give

$$\dot{V}_2 = -k_1 z_1^2 + z_2(z_1 + a_1 x_3 x_5 - T_L - J\dot{\alpha}_1). \tag{6.48}$$

Construct the virtual control law of this step as

$$\alpha_2 = \frac{1}{a_1 x_5}\left(-k_2 z_2 - z_1 + T_L + J\dot{\alpha}_1\right), \tag{6.49}$$

with $\dot{\alpha}_1 = -k_1(x_2 - \dot{x}_{1d}) + \ddot{x}_{1d}$ and $k_2 > 0$. Adding and subtracting α_2 in (6.48) gives

$$\dot{V}_2 = -k_1 z_1^2 - k_2 z_2^2 + a_1 x_5 z_2 z_3, \tag{6.50}$$

with $z_3 = x_3 - \alpha_2$.

Step 3: Computing the time derivative form of z_3, the following equation is available.

$$\dot{z}_3 = \dot{x}_3 - \dot{\alpha}_2 = b_1 x_4 - b_2 x_3 + b_3 \frac{x_3 x_6}{x_5} + x_2 x_6 - \dot{\alpha}_2. \tag{6.51}$$

Choose $V_3 = V_2 + \frac{1}{2} z_3^2$ and differentiating V_3 yields

$$\begin{aligned} \dot{V}_3 &= \dot{V}_2 + z_3 \dot{z}_3 \\ &= -k_1 z_1^2 - k_2 z_2^2 + z_3 \left(a_1 x_5 z_2 + b_1 x_4 - b_2 x_3 + b_3 \frac{x_3 x_6}{x_5} + x_2 x_6 - \dot{\alpha}_2 \right), \end{aligned} \tag{6.52}$$

with

$$\begin{aligned} \dot{\alpha}_2 = & \frac{1}{a_1 x_5}\Big(-\frac{k_2}{J} a_1 x_3 x_5 + \frac{k_2}{J} T_L + k_2 \ddot{x}_{1d} - k_1 k_2 x_2 - x_2 + \dot{x}_{1d} \\ & + J \dddot{x}_{1d} - k_1 a_1 x_3 x_5 + k_1 T_L + J k_1 \ddot{x}_{1d}\Big) + \frac{1}{a_1 x_5^2} [(d_1 x_5 + d_2 x_6) \\ & (k_2 z_2 + z_1 - T_L - J \ddot{x}_{1d} + J k_1 x_2 - J k_1 \dot{x}_{1d})]. \end{aligned} \tag{6.53}$$

And construct the virtual control of this step as

$$\alpha_3 = -\frac{1}{b_1} \left(k_3 z_3 + a_1 z_2 x_5 - b_2 x_3 + b_3 \frac{x_3 x_6}{x_5} + x_2 x_6 - \dot{\alpha}_2 \right), \tag{6.54}$$

with $k_3 > 0$.

Substituting (6.53) and (6.54) into (6.52) and introducing $z_4 = x_4 - \alpha_3$, one has

$$\dot{V}_3 = -k_1 z_1^2 - k_2 z_2^2 - k_3 z_3^2 + b_1 z_3 z_4. \tag{6.55}$$

Step 4: Computing the time derivative form of z_4, one has

$$\dot{z}_4 = \dot{x}_4 - \dot{\alpha}_3 = c_1 u_{qs} - c_2 x_4 + x_2 x_7 + c_3 \frac{x_3 x_7}{x_5} + c_4 x_3 - \dot{\alpha}_3. \tag{6.56}$$

Choose $V_4 = V_3 + \frac{1}{2} z_4^2$ and differentiate V_4 to obtain

$$\begin{aligned} \dot{V}_4 = \dot{V}_3 + z_4 \dot{z}_4 = & -k_1 z_1^2 - k_2 z_2^2 - k_3 z_3^2 + z_4 (b_1 z_3 + c_1 u_{qs} - c_2 x_4 \\ & + x_2 x_7 + c_3 \frac{x_3 x_7}{x_5} + c_4 x_3 - \dot{\alpha}_3). \end{aligned} \tag{6.57}$$

with

$$\dot{\alpha}_3 = -\frac{1}{b_1}(k_3\dot{z}_3 + a_1\dot{z}_2 x_5 + a_1 z_2\dot{x}_5 - b_2\dot{x}_3 + b_3(\frac{\dot{x}_3 x_6 x_5 + x_3\dot{x}_6 x_5 - \dot{x}_5 x_3 x_6}{x_5^2}) \\ +\dot{x}_2 x_6 + x_2\dot{x}_6 - \ddot{\alpha}_2),$$

and

$$\ddot{\alpha}_2 = \frac{1}{a_1^2 x_5^2}\{[-\frac{a_1}{J}k_1 x_3(d_1 x_5 + d_2 x_6) - \frac{a_1}{J}k_2 x_5(b_1 x_4 - b_1 x_3 + b_3\frac{x_3 x_6}{x_5} \\ +x_2 x_6) + k_2\dddot{x}_{1d} - k_1 k_2(\frac{1}{J}a_1 x_3 x_5 - \frac{T_L}{J}) + k_1 k_2\ddot{x}_{1d} - \frac{1}{J}a_1 x_3 x_5 \\ +\frac{T_L}{J} + \ddot{x}_{1d} + J\ddddot{x}_{1d} - k_1 a_1 x_3(d_1 x_5 + d_2 x_6) - k_1 a_1 x_5(b_1 x_4 \\ -b_1 x_3 + b_3\frac{x_3 x_6}{x_5} + x_2 x_6) + Jk_1\dddot{x}_{1d}]a_1 x_5 - (a_1 d_1 x_5 + a_1 d_2 x_6) \\ (-\frac{k_2}{J}a_1 x_3 x_5 + \frac{k_2}{J}T_L + k_2\ddot{x}_{1d} - k_1 k_2 x_2 + k_1 k_2\dot{x}_{1d} - x_2 + \dot{x}_{1d} + J\dddot{x}_{1d} \\ -k_1 a_1 x_3 x_5 + k_1 T_L + Jk_1\ddot{x}_{1d})\} + \frac{1}{a_1^2 x_5^4}\{[(d_1^2 x_5 + d_1 d_2 x_6 + d_2 b_1 x_7 \\ +d_2 d_3 x_5 - d_2 b_2 x_6 + b_3 d_2\frac{x_3^2}{x_5} + d_2 x_2 x_3)(k_2 z_2 + z_1 - T_L - J\ddot{x}_{1d} \\ +Jk_1 x_2 - Jk_1\dot{x}_{1d}) + (d_1 x_5 + d_2 x_6)(k_2\frac{1}{J}a_1 x_3 x_5 - \frac{T_L}{J}k_2 - k_2\dot{\alpha}_1 \\ +x_2 - \dot{x}_{1d} - J\dddot{x}_{1d} + a_1 k_1 x_3 x_5 - T_L k_1 - Jk_1\ddot{x}_{1d})]a_1 x_5^2 - (2a_1 x_5^2 d_1 \\ +2a_1 x_5 x_6 d_2)(d_1 x_5 + d_2 x_6)(k_2 z_2 + z_1 - T_L - J\ddot{x}_{1d} + Jk_1 x_2 - Jk_1\dot{x}_{1d})\}.$$

So we design the real control law u_{qs} as

$$u_{qs} = \frac{1}{c_1}(-k_4 z_4 - b_1 z_3 + c_2 x_4 - x_2 x_7 - c_3\frac{x_3 x_7}{x_5} - c_4 x_3 + \dot{\alpha}_3), \tag{6.58}$$

with $k_4 > 0$.

By use of (6.58), we can easily confirm that

$$\dot{V}_4 = -k_1 z_1^2 - k_2 z_2^2 - k_3 z_3^2 - k_4 z_4^2. \tag{6.59}$$

Step 5: Consider the second desired signal x_{5d} and introduce the tracking error $z_5 = x_5 - x_{5d}$. The Lyapunov candidate function V_5 is defined as $V_5 = V_4 + \frac{1}{2}z_5^2$, then its differential form can be computed as

$$\dot{V}_5 = \dot{V}_4 + z_5\dot{z}_5 = \dot{V}_4 + z_5(d_1 x_5 + d_2 x_6 - \dot{x}_{5d}). \tag{6.60}$$

Now, design the virtual control law of this step as

$$\alpha_4 = (-k_5 z_5 + \dot{x}_{5d} - d_1 x_5)/d_2, \tag{6.61}$$

with $k_5 > 0$.

Define $z_6 = x_6 - \alpha_4$. By using (6.61), (6.60) can be expressed as

$$\dot{V}_5 = \dot{V}_4 - k_5 z_5^2 + d_2 z_5 z_6. \tag{6.62}$$

Step 6: Differentiating z_6 gives

$$\dot{z}_6 = \dot{x}_6 - \dot{\alpha}_4 = b_1 x_7 + d_3 x_5 - b_2 x_6 + b_3 \frac{x_3^2}{x_5} + x_2 x_3 - \dot{\alpha}_4. \tag{6.63}$$

At present stage, consider $V_6 = V_5 + \frac{1}{2} z_6^2$. Clearly, its differential form is computed by

$$\dot{V}_6 = \dot{V}_5 + z_6 \dot{z}_6 = \dot{V}_4 - k_5 z_5^2 + z_6 (d_2 z_5 + b_1 x_7 + d_3 x_5 - b_2 x_6 + b_3 \frac{x_3^2}{x_5} + x_2 x_3 - \dot{\alpha}_4), \tag{6.64}$$

with

$$\dot{\alpha}_4 = \frac{1}{d_2} \left[-k_5 (d_1 x_5 + d_2 x_6 - \ddot{x}_{5d}) - d_1 \dot{x}_5 + \ddot{x}_{5d} \right],$$

and

$$\ddot{\alpha}_4 = \frac{1}{d_2} \left[-k_5 (d_1 \dot{x}_5 + d_2 \dot{x}_6 - \dddot{x}_{5d}) - d_1 (d_1 \dot{x}_5 + d_2 \dot{x}_6) + \dddot{x}_{5d} \right].$$

Design the virtual control law of this step as

$$\alpha_5 = -\frac{1}{b_1} \left(k_6 z_6 + d_2 z_5 + d_3 x_5 - b_2 x_6 + b_3 \frac{x_3^2}{x_5} + x_2 x_3 - \dot{\alpha}_4 \right), \tag{6.65}$$

with $k_6 > 0$.

Substituting (6.65) into (6.64) and defining $z_7 = x_7 - \alpha_5$, we can obtain

$$\dot{V}_6 = \dot{V}_4 - k_5 z_5^2 - k_6 z_6^2 + b_1 z_6 z_7. \tag{6.66}$$

Step 7: During this design step, the real control law u_{ds} will be designed. Differentiating z_7 gives

$$\dot{z}_7 = \dot{x}_7 - \dot{\alpha}_5 = c_1 u_{ds} - c_2 x_7 + c_3 \frac{x_3 x_4}{x_5} + x_2 x_4 - c_5 x_5 + c_4 x_6 - \dot{\alpha}_5, \tag{6.67}$$

where

$$\dot{\alpha}_5 = -\frac{1}{b_1}(k_6\dot{z}_6 + d_2\dot{z}_5 + d_3\dot{x}_5 - b_2\dot{x}_6 + b_3(\frac{2x_3\dot{x}_3x_5 - \dot{x}_5x_3^2}{x_5^2}) + \dot{x}_2x_3 + x_2\dot{x}_3 - \ddot{\alpha}_4).$$

Choose $V_7 = V_6 + \frac{1}{2}z_7^2$ and differentiate V_7 to have

$$\dot{V}_7 = \dot{V}_6 + z_7\dot{z}_7 = \dot{V}_4 - k_5z_5^2 - k_6z_6^2 + z_7(b_1z_6 + c_1u_{ds} - c_2x_7 + c_3\frac{x_3x_4}{x_5} + x_2x_4 - c_5x_5 + c_4x_6 - \dot{\alpha}_5). \quad (6.68)$$

Now design u_{ds} as

$$u_{ds} = \frac{1}{c_1}(-k_7z_7 - b_1z_6 + c_2x_7 - c_3\frac{x_3x_4}{x_5} - x_2x_4 + c_5x_5 - c_4x_6 + \dot{\alpha}_5), \quad (6.69)$$

with $k_7 > 0$.

Remark 6.8 Compared the fuzzy adaptive DSC (6.22) and (6.36) with the traditional backstepping controllers (6.58) and (6.69), it is obvious that the traditional controllers (6.58) and (6.69) are much more complicated than the DSC. Moreover, the number of terms in traditional controllers is much larger, which will bring online computation burdens and difficulties for practical applications. ♦

6.5 Simulation

Compared with traditional adaptive backstepping method, the proposed DSC control scheme is provided to demonstrate the benefits in this section. Nevertheless, the traditional backstepping design process is given. The simulation is run for IM with the parameters:

$$J = 0.0586\,\text{kg}\cdot\text{m}^2, \text{R}_\text{s} = 0.1, \text{R}_\text{r} = 0.15, \text{R}_\text{fe} = 30,$$
$$L_{1s} = L_{1r} = 0.0699\text{H}, L_m = 0.068\text{H}, n_p = 1.$$

The developed DSC is utilized for this system and the control parameters are selected as:

$$k_1 = 56, k_2 = 140, k_3 = 140, k_4 = 560, k_5 = 7000, k_6 = 140, k_7 = 280,$$
$$\epsilon_1 = \epsilon_2 = \epsilon_4 = \epsilon_5 = 0.000033, \epsilon_3 = 0.001, r_1 = r_2 = 0.05,$$
$$m_1 = m_2 = 0.02, \quad l_2 = l_3 = l_4 = l_6 = l_7 = 0.25.$$

The fuzzy membership functions are chosen as:

$$\mu_{F_i^1} = \exp\left[\frac{-(x+5)^2}{2}\right], \quad \mu_{F_i^2} = \exp\left[\frac{-(x+4)^2}{2}\right],$$
$$\mu_{F_i^3} = \exp\left[\frac{-(x+3)^2}{2}\right], \quad \mu_{F_i^4} = \exp\left[\frac{-(x+2)^2}{2}\right],$$
$$\mu_{F_i^5} = \exp\left[\frac{-(x+1)^2}{2}\right], \quad \mu_{F_i^6} = \exp\left[\frac{-(x-0)^2}{2}\right],$$
$$\mu_{F_i^7} = \exp\left[\frac{-(x-1)^2}{2}\right], \quad \mu_{F_i^8} = \exp\left[\frac{-(x-2)^2}{2}\right],$$
$$\mu_{F_i^9} = \exp\left[\frac{-(x-3)^2}{2}\right], \quad \mu_{F_i^{10}} = \exp\left[\frac{-(x-4)^2}{2}\right],$$
$$\mu_{F_i^{11}} = \exp\left[\frac{-(x-5)^2}{2}\right].$$

The control parameters for the traditional backstepping are chosen as same as the DSC in this chapter. And the simulations are implemented under zero initial conditions. The desired signals are selected as: $x_{1d} = 0.5\sin(t) + 0.3\sin(0.5t)$, $x_{5d} = 1$, the load torque $T_L = \begin{cases} 0.5, 0 \le t \le 15, \\ 1, t \ge 15. \end{cases}$

Figures 6.1, 6.2, 6.3, 6.4, 6.5, 6.6, 6.7, 6.8, 6.9, 6.10, 6.11, 6.12, 6.13, 6.14, 6.15, 6.16, 6.17, 6.18, 6.19 and 6.20 display the simulation results. Among them, Figs. 6.1, 6.3, 6.5, 6.7, 6.9, 6.11, 6.13, 6.15, 6.17 and 6.19 show the fuzzy adaptive DSC method and Figs. 6.2, 6.4, 6.6, 6.8, 6.10, 6.12, 6.14, 6.16, 6.18 and 6.20 present the traditional backstepping control scheme. Figures 6.1 and 6.2 display the tracking performance of x_{1d} and Figs. 6.5 and 6.6 show the error between x_1 and x_{1d}. Figures 6.3 and 6.4 show the reference signal x_5 and x_{5d}. Figures 6.7 and 6.8 present the error between x_5 and x_{5d}. Figures 6.9, 6.10, 6.11, 6.12, 6.13, 6.14, 6.15 and 6.16 show the curves of i_{qm}, i_{qs}, i_{dm}, i_{ds}, respectively. Figures 6.17, 6.18, 6.19 and 6.20 demonstrate the trajectories of u_{qs} and u_{ds}. From these two figures, it is obvious that the controllers are bounded. The above simulation results indicate that even under load torque disturbance and parameter uncertainties, the proposed controllers are able to track the desired signals quite well.

Remark 6.9 It can be observed from Figs. 6.1, 6.2, 6.3, 6.4, 6.5, 6.6, 6.7, 6.8, 6.9, 6.10, 6.11, 6.12, 6.13, 6.14, 6.15, 6.16, 6.17, 6.18, 6.19 and 6.20 that the fuzzy adaptive DSC approach developed in this chapter can achieve a satisfied control performance. The overshoots of the proposed controller u_{qs} is smaller and the tracking performance is better than the classical backstepping controllers. Moreover, the developed control algorithm can tackle down the issue of "explosion of complexity" emerged in classical backstepping process. ♦

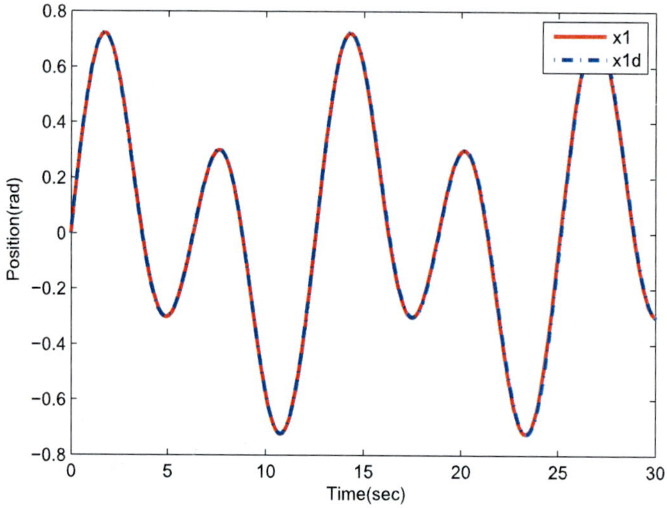

Fig. 6.1 Trajectories of the x_1 and x_{1d} for DSC

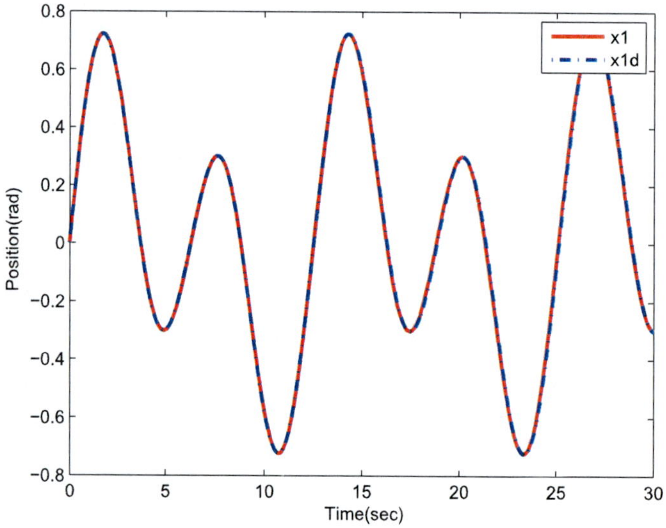

Fig. 6.2 Trajectories of the x_1 and x_{1d} for traditional backstepping

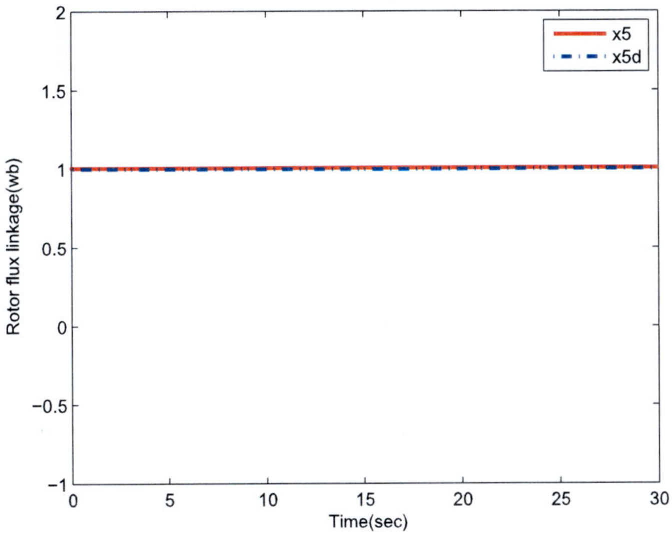

Fig. 6.3 Trajectories of the x_5 and x_{5d} for DSC

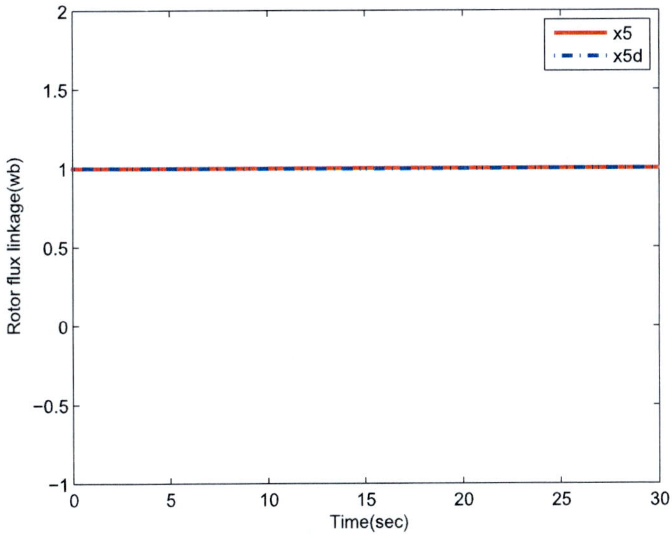

Fig. 6.4 Trajectories of the x_5 and x_{5d} for traditional backstepping

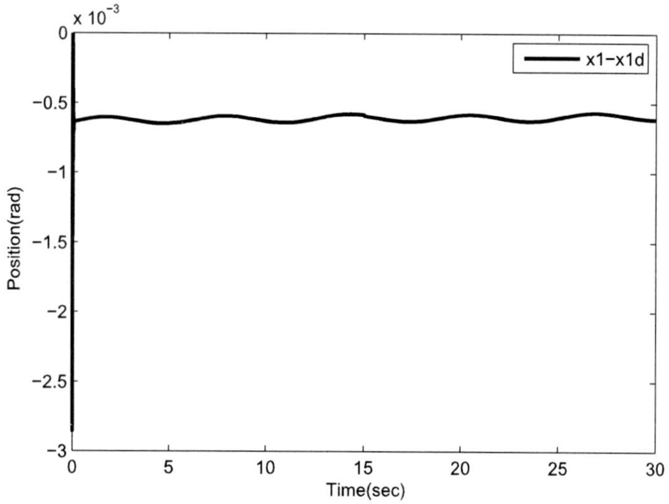

Fig. 6.5 The tracking error of x_1 and x_{1d} for DSC

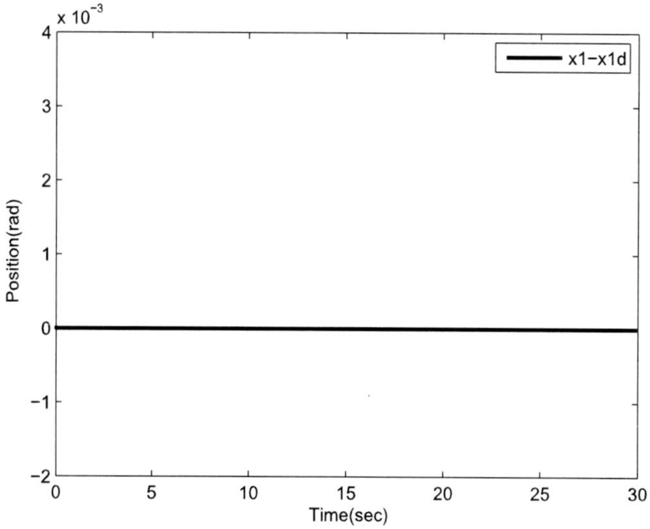

Fig. 6.6 The tracking error of x_1 and x_{1d} for traditional backstepping

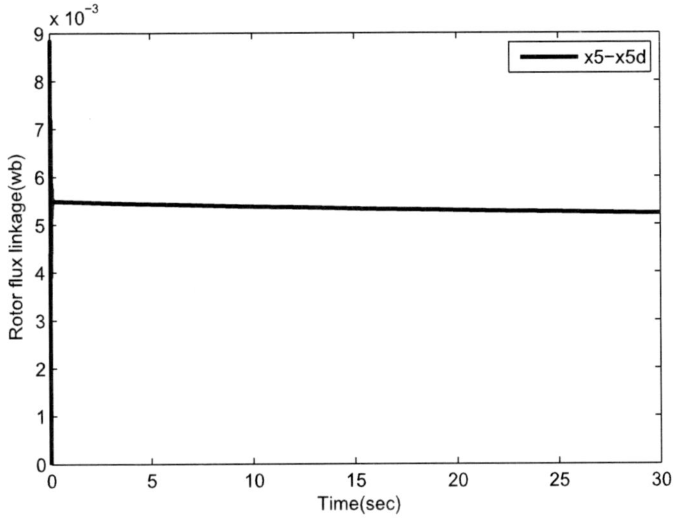

Fig. 6.7 The tracking error of x_5 and x_{5d} for DSC

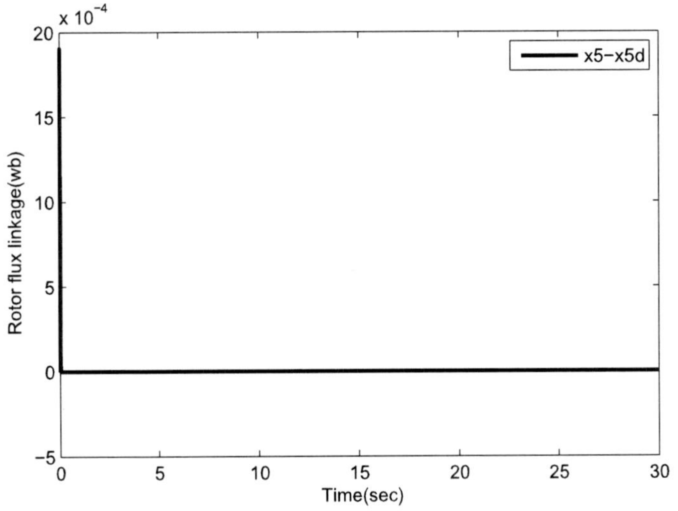

Fig. 6.8 The tracking error of x_5 and x_{5d} for traditional backstepping

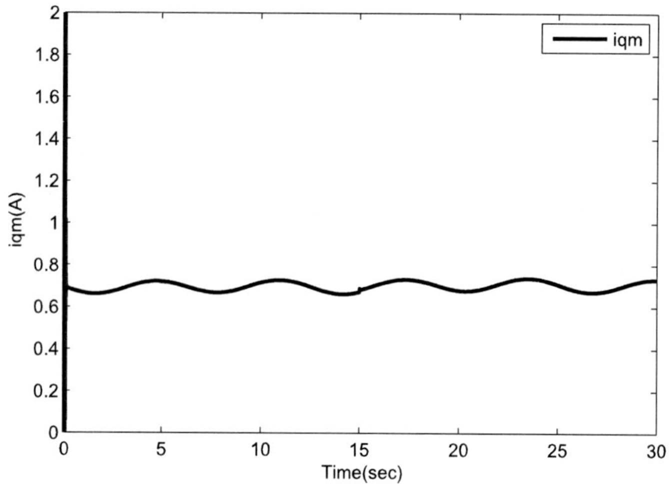

Fig. 6.9 Curve of the i_{qm} for DSC

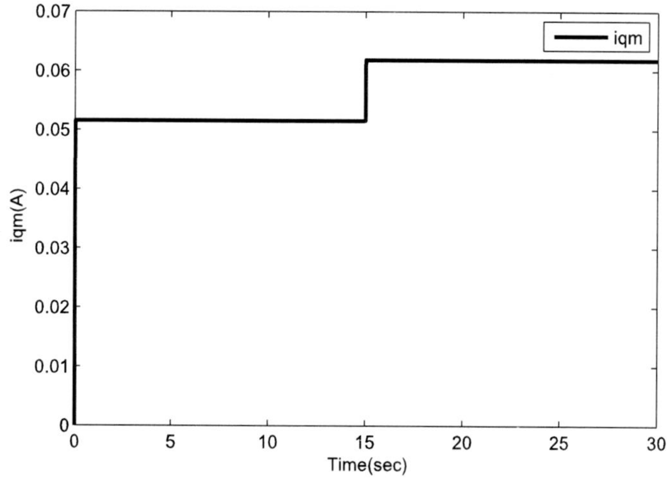

Fig. 6.10 Curve of the i_{qm} for traditional backstepping

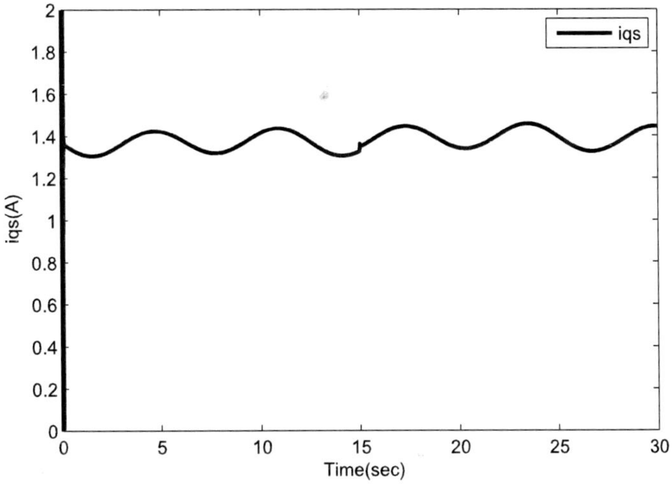

Fig. 6.11 Curve of the i_{qs} for DSC

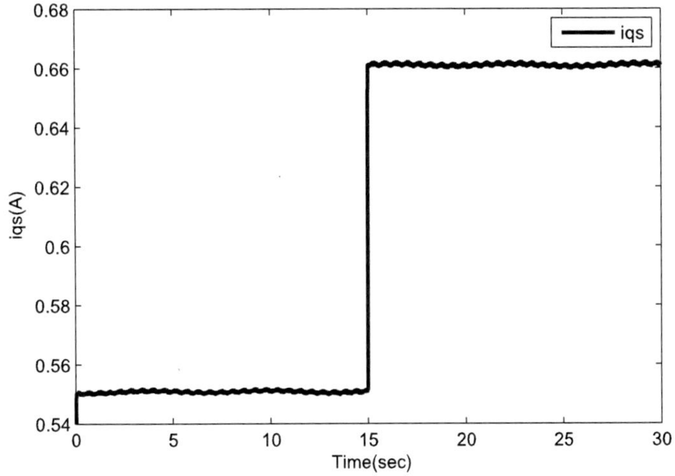

Fig. 6.12 Curve of the i_{qs} for traditional backstepping

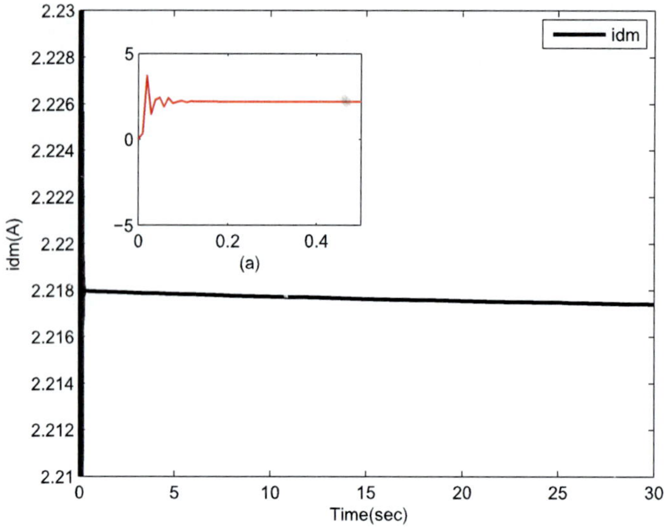

Fig. 6.13 Curve of the i_{dm} for DSC

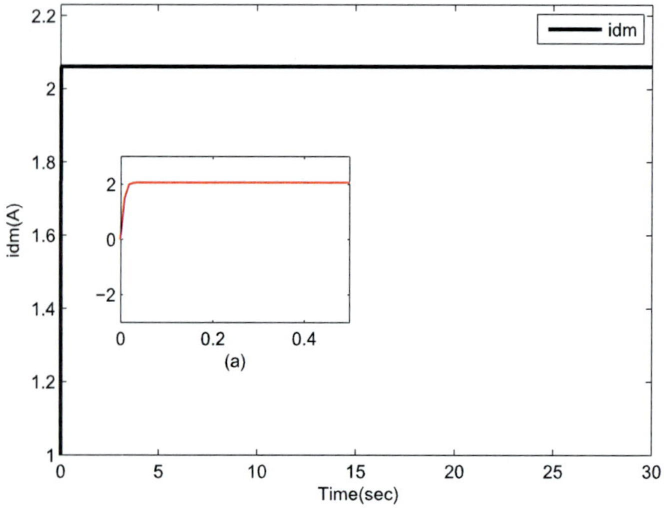

Fig. 6.14 Curve of the i_{dm} for traditional backstepping

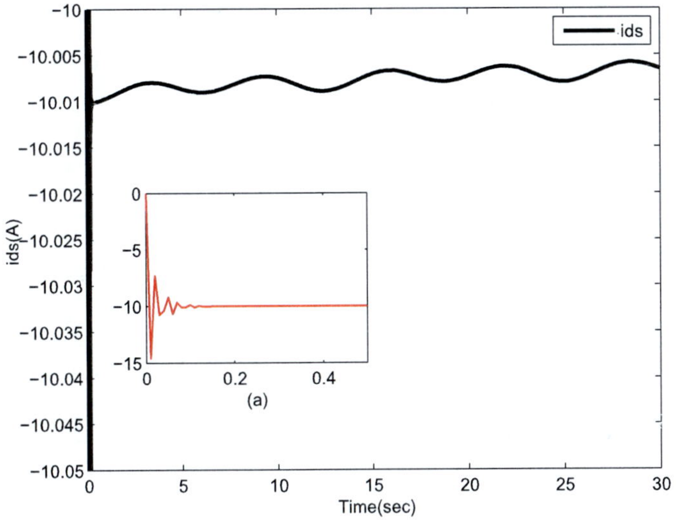

Fig. 6.15 Curve of the i_{ds} for DSC

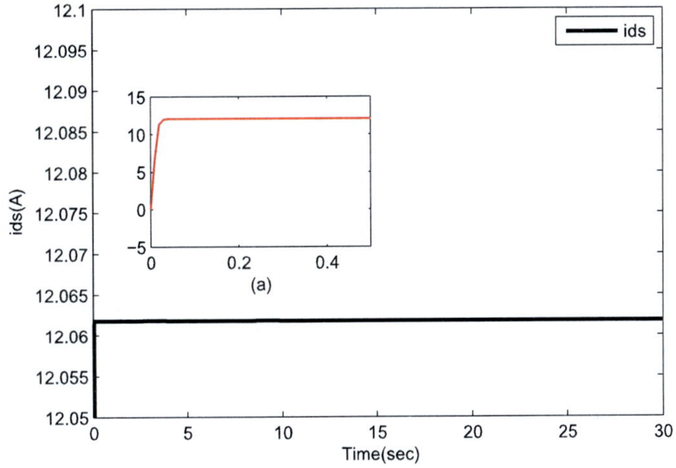

Fig. 6.16 Curve of the i_{ds} for traditional backstepping

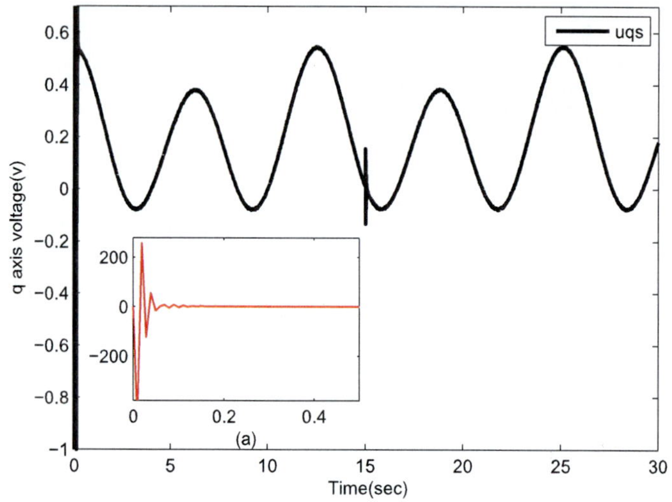

Fig. 6.17 Curve of the u_{qs} for DSC

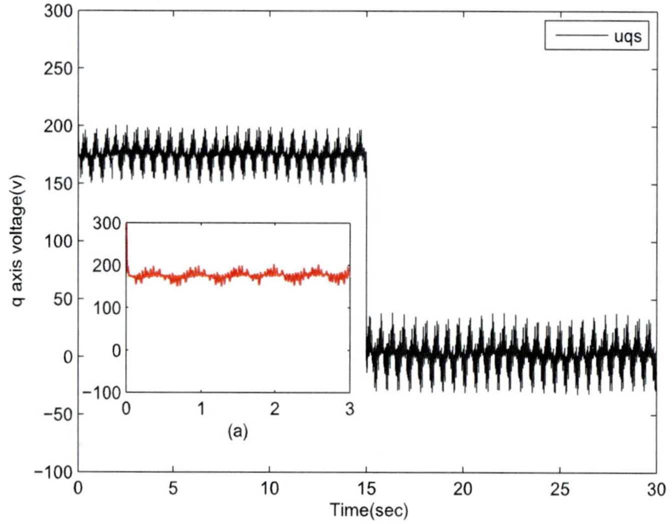

Fig. 6.18 Curve of the u_{qs} for traditional backstepping

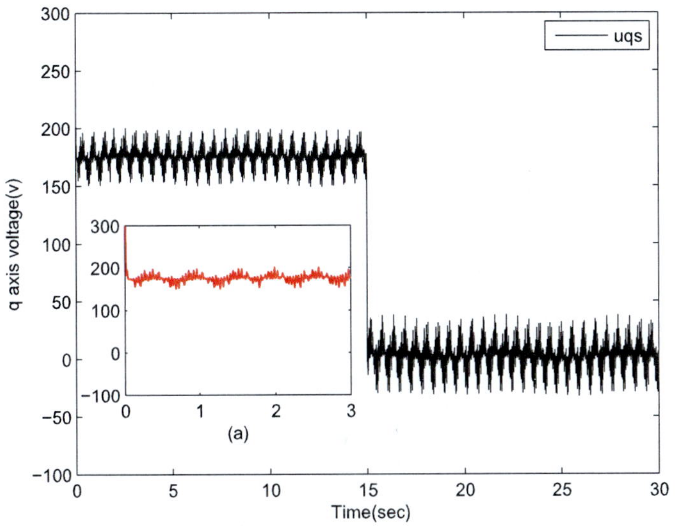

Fig. 6.19 Curve of the u_{ds} for DSC

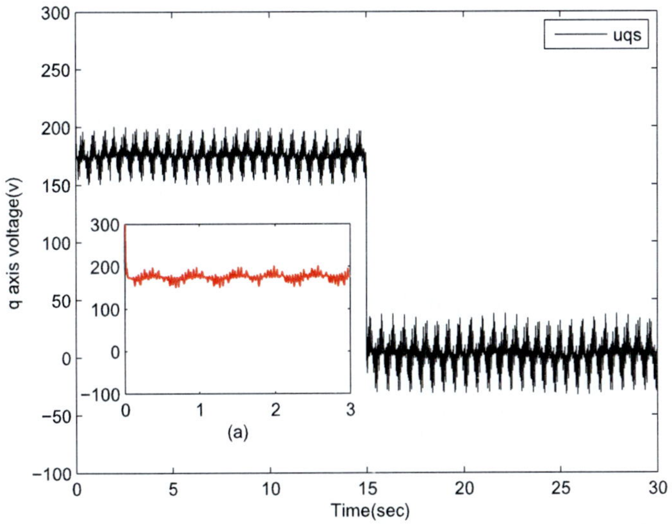

Fig. 6.20 Curve of the u_{ds} for traditional backstepping

6.6 Conclusion

The DSC procedure based on the adaptive fuzzy approximation is developed for IM with iron losses in electric vehicles in this chapter. The proposed controllers can solve two major problems of the classical backstepping. The present method can guarantee that the convergence of the tracking error falls into a small neighborhood of the origin and every closed-loop variable is bounded. Simulation results show the usefulness of the proposed approach including elimination of the influences from the load disturbance and parameter uncertainties.

References

1. Wang, H., Wang, D., Zhou, H.P.: Neural network based adaptive dynamic surface control for cooperative path following of marine surface vehicles via state and output feedback. Neurocomputing **133**, 170–178 (2014)
2. Ramirez, H.S., Ahmad, S., Zribi, M.: Dynamical feedback control of robotic manipulators with joint flexibility. IEEE Trans. Syst. Man Cybernet. **22**, 736–747 (1992)
3. Wai, R.J., Lin, K.M., Lin, C.Y.: Total sliding-mode speed control of field oriented induction motor servo drive. In: Proceedings of the 5th Asian Control Conference, Melbourne, Australia, pp. 1354–1361 (2004)
4. Yazdanpanah, R., Soltani, J., Arab Markadeh, G.R.: Nonlinear torque and stator flux controller for induction motor drive based on adaptive input-output feedback linearization and sliding mode control. Energy Convers. Manage. **49**(4), 541–550 (2008)
5. Marino, R., Peresada, S., Valigi, P.: Adaptive input-output linearizing control of induction motors. IEEE Trans. Automat. Contr. **38**(2), 208–221 (1993)
6. Prasad, D., Panigrahi, B.P., SenGupta, S.Z.: Digital simulation and hardware implementation of a simple scheme for direct torque control of induction motor. Energy Convers. Manage. **49**(4), 687–697 (2008)
7. Talaeizadeh, V., Kianinezhad, R., Seyfossadat, S.G.: Direct torque control of six-phase induction motors using three-phase matrix converter. Energy Convers. Manage. **51**(12), 2482–2491 (2010)
8. Krstic, M., Kanellakopoulos, I., Kokotovic, P.: Nonlinear and Adaptive Control Design. Wiley (1995)
9. Liu, X.P., Gu, G.X., Zhou, K.M.: Robust stabilization of MIMO nonlinear systems by backstepping. Automatica **35**(5), 987–992 (1999)
10. Yu, J.P., Chen, B., Yu, H.S.: Adaptive fuzzy tracking control for the chaotic permanent magnet synchronous motor drive system via backstepping. Nonlinear Analysis: Real World Applications **12**(1), 671–681 (2011)
11. Yu, J.P., Chen, B., Yu, H.S.: Position tracking control of induction motors via adaptive fuzzy backstepping. Energy Convers. Manage. **51**(11), 2345–2352 (2010)
12. Yu, W.S., Chen, H.S.: Interval type-2 fuzzy adaptive tracking control design for PMDC motor with the sector dead-zones. Inf. Sci. **288**, 108–134 (2014)
13. Zhou, Z.H., Zhao, J.W., Cao, F.L.: A novel approach for fault diagnosis of induction motor with invariant character vectors. Inf. Sci. **281**, 496–506 (2014)
14. Li, K., Zhang, C.H., Cui, N.X.: Vector control of induction motor for electric vehicles considering iron losses and its energy optimization strategy. Control Theory Appl. **24**(6), 959–963 (2007)
15. Pei, W.H., Zhang, C.H., Li, K., Cui, N.X.: Hamilton system modeling and passive control for induction motor of electric vehicles by considering iron losses. Control Theory Appl. **28**(6), 869–873 (2011)
16. Wang, L.X., Mendel, J.M.: Fuzzy basis functions, universal approximation, and orthogonal least squares learning. IEEE Trans. Neural Netw. **3**(5), 807–814 (1992)

Part II
Permanent Magnet Synchronous Motor (PMSM)

Chapter 7
Adaptive Fuzzy Tracking Control for a PMSM via Backstepping Approach

In this chapter, a speed tracking control is proposed for permanent magnet synchronous motor (PMSM) with parameter uncertainties [1, 2] and load torque disturbance. The nonlinearities are approximated by fuzzy logic systems and the controller is constructed by adaptive backstepping technique. The proposed controller can ensure that the position tracking error converges to a small neighborhood of the origin [3]. The simulation results show that the proposed control scheme can keep up with the desired signal well under the condition of parameter uncertainty and load torque disturbance without singularity and overparameterization.

7.1 Introduction

Permanent magnet synchronous motor (PMSM) has attracted great interest in industrial applications requiring dynamic performance due to its high speed [4], high efficiency, high power density and large torque to inertia ratio. However, it is still a challenging problem to control PMSM to obtain ideal dynamic performance because the motor dynamic model of PMSM is nonlinear and multivariable, the model parameters such as the stator resistance and the friction coefficient are also not exactly known. In recent years, the control of PMSM drivers has received extensive attention and some advanced control techniques, such as adaptive control [5, 6], backstepping principles [7–9] and fuzzy logic control [10–12], are used to solve the problems of speed or position tracking control of the PMSM.

In this chapter, the nonlinearities are approximated by fuzzy logic systems [13–15], and the adaptive fuzzy controllers are constructed via backstepping. The designed controller can track the reference signal quite well even the existence of the parameter uncertainties and load torque disturbance. Compared with the exist-

J. Yu et al., *Intelligent Backstepping Control for the Alternating-Current Drive Systems*, Studies in Systems, Decision and Control 349,
https://doi.org/10.1007/978-3-030-67723-7_7

ing based controller design schemes, the proposed method is very simple and the proposed controller has a simple structure.

7.2 Mathematical Model of the PMSM Drive System and Preliminaries

This section will introduce some preliminary knowledge of PMSM. The following assumptions are made in the derivation of the mathematical model of a PMSM [16]:

Assumption 7.1 [16] Saturation and iron losses are neglected although it can be taken into account by parameter changes.

Assumption 7.2 [16] The back emf is sinusoidal.

From (1.16), the dynamic model of a PMSM can be described by the following differential equations:

$$\begin{aligned} \dot{x}_1 &= \frac{a_1}{J}x_2 + \frac{a_2}{J}x_2x_3 - \frac{B}{J}x_1 - \frac{T_L}{J}, \\ \dot{x}_2 &= b_1x_2 + b_2x_1x_3 + b_3x_1 + b_4u_q, \\ \dot{x}_3 &= c_1x_3 + c_2x_1x_2 + c_3u_d, \end{aligned} \tag{7.1}$$

where x_1 denotes rotor angular velocity, x_2 and x_3 stand for the $d - q$ axis currents, u_d and u_q are the $d - q$ axis voltages, J means the rotor moment of inertia, B is the viscous friction coefficient, T_L is the load torque. The control objective is to design an adaptive fuzzy controller such that the state variable x_1 tracks the given reference signal x_d and all signals of the resulting closed-loop system are uniformly ultimately bounded.

Lemma 7.1 *[17] Let $f(x)$ be a continuous function defined on a compact set Ω. Then for any scalar $\varepsilon > 0$, there exists a fuzzy logic system $W^T S(x)$ such that*

$$\sup_{x\in\Omega} \left| f(x) - W^T S(x) \right| \leq \varepsilon,$$

where $W = [W_1, \ldots, W_N]^T$ is the ideal constant weight vector, and $S(x) = [p_1(x), p_2(x), \ldots, p_N(x)]^T / \sum_{i=1}^{N} p_i(x)$ is the basis function vector, with $N > 1$ being the number of the fuzzy rules and p_i are chosen as Gaussian functions, i.e., for $i = 1, 2, \ldots, N$, $p_i(x) = \exp[\frac{-(x-\mu_i)^T(x-\mu_i)}{\eta_i^2}]$ where $\mu_i = [\mu_{i1}, \mu_{i2}, \ldots, \mu_{in}]^T$ is the center vector, and η_i is the width of the Gaussian function.

7.3 Adaptive Fuzzy Controller with the Backstepping Technique

For the system (7.1), the backstepping design procedure contains 3 steps. A virtual control function α_1 will be constructed by using an appropriate Lyapunov function V. At the last step, a real controller is constructed to control the system. Next, we will give the procedure of the backstepping design.

Step 1: For the reference signal x_d, define the tracking error variable as $z_1 = x_1 - x_d$. From the first subsystem of (7.1), the error dynamic system is given by $\dot{z}_1 = \frac{a_1}{J}x_2 + \frac{a_2}{J}x_2x_3 - \frac{B}{J}x_1 - \frac{T_L}{J} - \dot{x}_d$.

Choose the Lyapunov function candidate as $V_1 = \frac{J}{2}z_1^2$, then the time derivative of V_1 is given by

$$\dot{V}_1 = Jz_1\dot{z}_1 = z_1(a_1x_2 + a_2x_2x_3 - Bx_1 - T_L - J\dot{x}_d). \tag{7.2}$$

Due to the parameters B, T_L and J are unknown, they cannot be used to construct the control signal. Thus, let $\hat{B}$, $\hat{T}_L$ and $\hat{J}$ be their estimations of B, T_L and J, respectively. The corresponding adaptation laws will be determined later. Now, construct the virtual control law α_1 as

$$\alpha_1(Z_1) = \frac{1}{a_1}(-k_1z_1 + \hat{B}x_1 + \hat{T}_L + \hat{J}\dot{x}_d), \tag{7.3}$$

where $k_1 > 0$ is a design parameter and $Z_1 = [x_1, x_d, \dot{x}_d, \hat{B}, \hat{T}_L, \hat{J}]^T$. Defining $z_2 = x_2 - \alpha_1$ and substituting (7.3) into (7.2) yield

$$\dot{V}_1 = -k_1z_1^2 + a_1z_1z_2 + a_2z_1x_2x_3 + z_1(\hat{B} - B)x_1 + z_1(\hat{T}_L - T_L) + z_1(\hat{J} - J)\dot{x}_d.$$

Step 2: Differentiating z_2 and using the second subsystem of (7.1) gives

$$\dot{z}_2 = \dot{x}_2 - \dot{\alpha}_1 = b_1x_2 + b_2x_1x_3 + b_3x_1 + b_4u_q - \dot{\alpha}_1. \tag{7.4}$$

Now, choose the Lyapunov function candidate as $V_2 = V_1 + \frac{1}{2}z_2^2$. Obviously, the time derivative of V_2 is showed by

$$\begin{aligned}\dot{V}_2 = \dot{V}_1 + z_2\dot{z}_2 &= -k_1z_1^2 + a_2z_1x_2x_3 + z_1(\hat{B} - B)x_1 + z_1(\hat{T}_L - T_L)\\ &+z_1(\hat{J} - J)\dot{x}_d + z_2(f_2 + b_4u_q),\end{aligned} \tag{7.5}$$

where

$$\dot{\alpha}_1 = \frac{\partial\alpha_1}{\partial x_1}\dot{x}_1 + \sum_{i=0}^{1}\frac{\partial\alpha_1}{\partial x_d^{(i)}}x_d^{(i+1)} + \frac{\partial\alpha_1}{\partial\hat{B}}\dot{\hat{B}} + \frac{\partial\alpha_1}{\partial\hat{T}_L}\dot{\hat{T}}_L + \frac{\partial\alpha_1}{\partial\hat{J}}\dot{\hat{J}}$$

$$= \frac{\partial \alpha_1}{\partial x_1}(\frac{a_1}{J}x_2 + \frac{a_2}{J}x_2x_3 - \frac{B}{J}x_1 - \frac{T_L}{J}) + \sum_{i=0}^{1} \frac{\partial \alpha_1}{\partial x_d^{(i)}} x_d^{(i+1)}$$
$$+ \frac{\partial \alpha_1}{\partial \hat{B}}\dot{\hat{B}} + \frac{\partial \alpha_1}{\partial \hat{T}_L}\dot{\hat{T}}_L + \frac{\partial \alpha_1}{\partial \hat{J}}\dot{\hat{J}},$$
$$f_2(Z_2) = a_1 z_1 + b_1 x_2 + b_2 x_1 x_3 + b_3 x_1 - \dot{\alpha}_1,$$
$$Z_2 = [x_1, x_2, x_3, x_d, \dot{x}_d, \ddot{x}_d, \hat{B}, \hat{T}_L, \hat{J}]^T.$$

Apparently, there are two nonlinear terms in (7.5), i.e., $a_2 z_1 x_2 x_3$ and f_2, wherein, f_2 contains the derivative of $\dot{\alpha}_1$. This will make the classical adaptive backstepping design become very complex and troubled, and the designed control law u_q will have a complex structure. To avoid this trouble in design procedure and simplify the control signal structure, we will apply the fuzzy logic system to approximate the nonlinear function f_2. As shown later, the design procedure of u_q becomes simple and u_q has the simple structure. According to Lemma 7.1, for any given $\varepsilon_2 > 0$, there exists a fuzzy logic system $W_2^T S_2(Z_2)$ such that

$$f_2(Z_2) = W_2^T S_2(Z_2) + \delta_2(Z_2),$$

with $\delta_2(Z_2)$ being the approximation error and satisfying $|\delta_2| \leq \varepsilon_2$. Consequently, a simple computing method produces the following inequality.

$$z_2 f_2 = z_2\left(W_2^T S_2 + \delta_2\right) \leq \frac{1}{2l_2^2} z_2^2 \|W_2\|^2 S_2^2 + \frac{1}{2} l_2^2 + \frac{1}{2} z_2^2 + \frac{1}{2}\varepsilon_2^2. \tag{7.6}$$

It follows immediately from substituting (7.6) into (7.5) that

$$\dot{V}_2 \leq -\sum_{i=1}^{2} k_i z_i^2 + a_2 z_1 x_2 x_3 + z_1(\hat{B} - B)x_1 + z_1(\hat{T}_L - T_L)$$
$$+ z_1(\hat{J} - J)\dot{x}_d + \frac{1}{2l_2^2} z_2^2 \|W_2\|^2 S_2^2 + \frac{1}{2} l_2^2 + \frac{1}{2} z_2^2 + \frac{1}{2}\varepsilon_2^2 + z_2 b_4 u_q.$$

The control input u_q is designed as

$$u_q = \frac{1}{b_4}(-k_2 z_2 - \frac{1}{2} z_2 - \frac{1}{2l_2^2} z_2 \hat{\theta} S_2^2), \tag{7.7}$$

where $\hat{\theta}$ is the estimation of the unknown constant θ which will be specified later. Using the equality (7.7), thus the derivative of V_2 becomes as

$$\begin{aligned}\dot{V}_2 \le & -\sum_{i=1}^{2} k_i z_i^2 + a_2 z_1 x_2 x_3 + z_1(\hat{B} - B)x_1 + z_1(\hat{T}_L - T_L) \\ & + z_1(\hat{J} - J)\dot{x}_d + \frac{1}{2l_2^2} z_2^2 (\|W_2\|^2 - \hat{\theta}) S_2^2 + \frac{1}{2} l_2^2 + \frac{1}{2}\varepsilon_2^2.\end{aligned}$$

Step 3: At this step, we will construct the control law u_d. To this end, define $z_3 = x_3$ and choose the following Lyapunov function candidate as $V_3 = V_2 + \frac{1}{2} z_3^2$. Then the derivative of V_3 is given by

$$\begin{aligned}\dot{V}_3 = & \dot{V}_2 + z_3 \dot{z}_3 \\ \le & -\sum_{i=1}^{2} k_i z_i^2 + a_2 z_1 x_2 x_3 + z_1(\hat{B} - B)x_1 + z_1(\hat{T}_L - T_L) \\ & + z_1(\hat{J} - J)\dot{x}_d + \frac{1}{2l_2^2} z_2^2 (\|W_2\|^2 - \hat{\theta}) S_2^2 + \frac{1}{2} l_2^2 + \frac{1}{2}\varepsilon_2^2 \\ & + z_3 \left(f_3(Z_3) + c_3 u_d \right),\end{aligned} \tag{7.8}$$

where $f_3(Z_3) = a_2 z_1 x_2 + c_1 x_3 + c_2 x_1 x_2$ and $Z_3 = [x_1, x_2, x_3, x_d]^T$. Similarly, by Lemma 7.1 the fuzzy logic system $W_3^T S_3(Z_3)$ is utilized to approximate the nonlinear function f_3 such that for given $\varepsilon_3 > 0$,

$$z_3 f_3 \le \frac{1}{2l_3^2} z_3^2 \|W_3\|^2 S_3^2 + \frac{1}{2} l_3^2 + \frac{1}{2} z_3^2 + \frac{1}{2}\varepsilon_3^2. \tag{7.9}$$

Substituting (7.9) into (7.8) gives

$$\begin{aligned}\dot{V}_3 = & \dot{V}_2 + z_3 \dot{z}_3 \\ \le & -\sum_{i=1}^{2} k_i z_i^2 + \frac{1}{2l_2^2} z_2^2 (\|W_2\|^2 - \hat{\theta}) S_2^2 + \sum_{i=2}^{3} \frac{1}{2}(l_i^2 + \varepsilon_i^2) \\ & + z_1(\hat{B} - B)x_1 + z_1(\hat{T}_L - T_L) + z_1(\hat{J} - J)\dot{x}_d \\ & + \frac{1}{2l_3^2} z_3^2 \|W_3\|^2 S_3^2 + \frac{1}{2} z_3^2 + c_3 z_3 u_d.\end{aligned} \tag{7.10}$$

Now design u_d as

$$u_d = \frac{-1}{c_3}(k_3 z_3 + \frac{1}{2} z_3 + \frac{1}{2l_3^2} z_3 \hat{\theta} S_3^2). \tag{7.11}$$

Then, define $\theta = \max\{\|W_2\|^2, \|W_3\|^2\}$. Then, combining (7.10) with (7.11) results in

$$\begin{aligned}\dot{V}_4 \leq & -\sum_{i=1}^{3} k_i z_i^2 + \sum_{i=2}^{3} \frac{1}{2}(l_i^2 + \varepsilon_i^2) + z_1(\hat{B} - B)x_1 + z_1(\hat{T}_L - T_L) \\ & + z_1(\hat{J} - J)\dot{x}_d + \sum_{i=2}^{3} \frac{1}{2l_i^2} z_i^2 \left(\|W_i\|^2 - \hat{\theta}\right) S_i^T(Z_i) S_i(Z_i) \\ \leq & -\sum_{i=1}^{3} k_i z_i^2 + \sum_{i=2}^{3} \frac{1}{2}(l_i^2 + \varepsilon_i^2) + z_1(\hat{B} - B)x_1 + z_1(\hat{T}_L - T_L) \\ & + z_1(\hat{J} - J)\dot{x}_d + \sum_{i=2}^{3} \frac{1}{2l_i^2} z_i^2 S_i^T(Z_i) S_i(Z_i) \left(\theta - \hat{\theta}\right). \end{aligned} \tag{7.12}$$

At the present stage, to estimate the unknown constants B, T_L, J and θ, define the adaptive variables as follows

$$\begin{aligned}\tilde{T}_L &= \hat{T}_L - T_L, \\ \tilde{B} &= \hat{B} - B, \\ \tilde{J} &= \hat{J} - J, \\ \tilde{\theta} &= \hat{\theta} - \theta. \end{aligned} \tag{7.13}$$

In order to determine the corresponding adaptation laws, the following Lyapunov function candidates are selected as:

$$V = V_3 + \frac{1}{2r_1}\tilde{T}_L^2 + \frac{1}{2r_2}\tilde{B}^2 + \frac{1}{2r_3}\tilde{J}^2 + \frac{1}{2r_4}\tilde{\theta}^2, \tag{7.14}$$

where $r_i, i = 1, 2, 3, 4$ are positive constants. By differentiating V and taking (7.12)–(7.14) into account, one has

$$\begin{aligned}\dot{V} \leq & -\sum_{i=1}^{3} k_i z_i^2 + \sum_{i=2}^{3} \frac{1}{2}(l_i^2 + \varepsilon_i^2) + z_1 \tilde{B} x_1 + z_1 \tilde{T}_L + z_1 \tilde{J} \dot{x}_d \\ & -\sum_{i=2}^{3} \frac{1}{2l_i^2} z_i^2 \tilde{\theta} S_i^T(Z_i) S_i(Z_i) + \frac{1}{r_1}\tilde{T}_L \dot{\hat{T}}_L + \frac{1}{r_2}\tilde{B}\dot{\hat{B}} + \frac{1}{r_3}\tilde{J}\dot{\hat{J}} + \frac{1}{r_4}\tilde{\theta}\dot{\hat{\theta}} \\ = & -\sum_{i=1}^{3} k_i z_i^2 + \sum_{i=2}^{3} \frac{1}{2}(l_i^2 + \varepsilon_i^2) + \frac{1}{r_1}\tilde{T}_L \left(r_1 z_1 + \dot{\hat{T}}_L\right) + \frac{1}{r_2}\tilde{B}\left(r_2 z_1 x_1 + \dot{\hat{B}}\right) \\ & + \frac{1}{r_3}\tilde{J}\left(r_3 z_1 \dot{x}_d + \dot{\hat{J}}\right) + \frac{1}{r_4}\tilde{\theta}\left[-\sum_{i=2}^{3} \frac{r_4}{2l_i^2} z_i^2 S_i^T(Z_i) S_i(Z_i) + \dot{\hat{\theta}}\right]. \end{aligned} \tag{7.15}$$

According to (7.15), the corresponding adaptive laws are selected as follows:

$$\begin{aligned}
\dot{\hat{T}}_L &= -r_1 z_1 - m_1 \hat{T}_L, \\
\dot{\hat{B}} &= -r_2 z_1 x_1 - m_2 \hat{B}, \\
\dot{\hat{J}} &= -r_3 z_1 \dot{x}_d - m_3 \hat{J}, \\
\dot{\hat{\theta}} &= \sum_{i=2}^{3} \frac{r_4}{2l_i^2} z_i^2 S_i^T(Z_i) S_i(Z_i) - m_4 \hat{\theta},
\end{aligned} \tag{7.16}$$

where m_i for $i = 1, 2, 3, 4$ and l_i for $i = 2, 3$ are positive constants.

Theorem 7.2 *Consider system (7.1) and reference signal x_{1d}. If the virtual control signals are constructed as in (7.3), the adaptive law is designed as in (7.16), then we choose the adaptive fuzzy controllers (7.7) and (7.11) such that the resulting tracking errors converge to the small neighborhood of the origin. Also, all closed-loop signals of the controlled system are bounded.*

Proof In this section, the stability analysis of the resulting closed-loop system will be addressed. Substituting (7.16) into (7.15) yields

$$\dot{V} \le -\sum_{i=1}^{3} k_i z_i^2 + \sum_{i=2}^{3} \frac{1}{2}(l_i^2 + \varepsilon_i^2) - \frac{m_1}{r_1}\tilde{T}_L \hat{T} - \frac{m_2}{r_2}\tilde{B}\hat{B} - \frac{m_3}{r_3}\tilde{J}\hat{J} - \frac{m_4}{r_4}\tilde{\theta}\hat{\theta}. \tag{7.17}$$

For the term $-\tilde{T}_L \hat{T}$, one has

$$-\tilde{T}_L \hat{T}_L = -\tilde{T}_L(\tilde{T}_L + T_L) \le -\frac{1}{2}\tilde{T}_L^2 + \frac{1}{2}T_L^2.$$

Similarly, we have

$$\begin{aligned}
-\tilde{B}\hat{B} &\le -\frac{1}{2}\tilde{B}^2 + \frac{1}{2}B^2, \\
-\tilde{J}\hat{J} &\le -\frac{1}{2}\tilde{J}^2 + \frac{1}{2}J^2, \\
-\tilde{\theta}\hat{\theta} &\le -\frac{1}{2}\tilde{\theta}^2 + \frac{1}{2}\theta^2.
\end{aligned}$$

Consequently, by using these inequalities (7.17) can be rewritten in the following form

$$\begin{aligned}
\dot{V} \le& -\sum_{i=1}^{3} k_i z_i^2 - \frac{m_1}{2r_1}\tilde{T}_L^2 - \frac{m_2}{2r_2}\tilde{B}^2 - \frac{m_3}{2r_3}\tilde{J}^2 - \frac{m_4}{2r_4}\tilde{\theta}^2 \\
&+ \sum_{i=2}^{3} \frac{1}{2}(l_i^2 + \varepsilon_i^2) + \frac{m_1}{2r_1}T_L^2 + \frac{m_2}{2r_2}B^2 + \frac{m_3}{2r_3}J^2 + \frac{m_4}{2r_4}\theta^2 \\
\le& -a_0 V + b_0,
\end{aligned} \tag{7.18}$$

where $a_0 = \min\left\{\frac{2k_1}{J}, 2k_2, 2k_3, m_1, m_2, m_3, m_4\right\}$ and $b_0 = \sum_{i=2}^{3} \frac{1}{2}(l_i^2 + \varepsilon_i^2) + \frac{m_1}{2r_1} T_L^2 + \frac{m_2}{2r_2} B^2 + \frac{m_3}{2r_3} J^2 + \frac{m_4}{2r_4} \theta^2$. Furthermore, (7.18) implies that

$$V(t) \leq (V(t_0) - \frac{b_0}{a_0})e^{-a_0(t-t_0)} + \frac{b_0}{a_0} \leq V(t_0) + \frac{b_0}{a_0}, \forall t \geqslant t_0. \tag{7.19}$$

As a result, all $z_i (i = 1, 2, 3)$, $\tilde{T}_L$, $\tilde{B}$, $\tilde{J}$ and $\tilde{\theta}$ belong to the compact set

$$\Omega = \left\{(z_i, \tilde{T}_L, \tilde{B}, \tilde{J}, \tilde{\theta}) | V \leq V(t_0) + \frac{b_0}{a_0}, \forall t \geqslant t_0\right\}.$$

Namely, all signals in the closed-loop system are bounded. Especially, from (7.19) we have

$$\lim_{t\to\infty} z_1^2 \leq \frac{2b_0}{a_0}.$$

From the definitions of a_0 and b_0, it is clear that to get a small tracking error we can take r_i large and l_i and ε_i small enough after giving the parameters k_i and m_i. ■

7.4 A Comparison with the Conventional Backstepping Design

7.4.1 Conventional Backstepping Design

The PMSM controller based on conventional backstepping technique is introduced briefly here. The controllers based on conventional backstepping according to [9] are shown as follows

$$\begin{aligned} u_q = & \frac{1}{b_4}(-k_2 z_2 + b_4(\frac{\dot{\hat{J}}}{a_1}(-k_1 z_1 + \hat{F} x_1 + \hat{\Gamma} + \dot{x}_d) \\ & + \frac{\hat{J}}{a_1}(\dot{\hat{F}} x_1 \dot{\hat{\Gamma}} \ddot{x}_d + k_1 \dot{x}_d) + (\hat{F} - k_1)(x_2 + \frac{a_2}{a_1} x_2 x_3) \\ & - \frac{\hat{J}}{a_1}(\hat{F} - k_1)(\hat{F} x_1 + \hat{\Gamma}) - (b_1 x_2 + b_2 x_1 x_3 + b_3 x_1))), \\ u_d = & \frac{1}{c_3}(-k_3 z_3 - c_1 x_3 - c_2 x_1 x_2 - \frac{a_2}{\hat{J}} c_3 z_1 x_2). \end{aligned}$$

Compared with the traditional backstepping controller, we can see that the structure of the controller based on adaptive fuzzy backstepping is simple. And the simulation results illustrate the effectiveness of the presented method in this article.

7.4.2 Simulation

In order to verify the effectiveness of the proposed results, the simulation will be done for the PMSM with the parameters:

$$\begin{aligned} &J = 0.00379\,\mathrm{Kg}\cdot\mathrm{m}^2, \mathrm{R_s} = 0.68\,\Omega, \mathrm{L_d} = 0.00315\,\mathrm{H}, \\ &L_q = 0.00285\,\mathrm{H}, \mathrm{B} = 0.001158\,\mathrm{N}\cdot\mathrm{m}/(\mathrm{rad/s}), \Phi = 0.1245\,\mathrm{Wb}, \mathrm{n_p} = 3, \end{aligned}$$

Then, the proposed adaptive fuzzy controllers are used to control this PMSM. Given the reference signal as $x_d = 30$ and the control parameters are chosen as follows:

$$\begin{aligned} &k_1 = 2.5, \quad k_2 = k_3 = 50, \quad r_1 = r_2 = r_3 = r_4 = 2.5, \\ &m_1 = m_2 = m_3 = m_4 = 0.0005, \quad l_2 = l_3 = 5. \end{aligned}$$

The fuzzy membership functions are chosen as:

$$\begin{aligned} &\mu_{F_i^1} = \exp\left[\frac{-(x+5)^2}{2}\right], \quad \mu_{F_i^2} = \exp\left[\frac{-(x+4)^2}{2}\right], \\ &\mu_{F_i^3} = \exp\left[\frac{-(x+3)^2}{2}\right], \quad \mu_{F_i^4} = \exp\left[\frac{-(x+2)^2}{2}\right], \\ &\mu_{F_i^5} = \exp\left[\frac{-(x+1)^2}{2}\right], \quad \mu_{F_i^6} = \exp\left[\frac{-(x-0)^2}{2}\right], \\ &\mu_{F_i^7} = \exp\left[\frac{-(x-1)^2}{2}\right], \quad \mu_{F_i^8} = \exp\left[\frac{-(x-2)^2}{2}\right], \\ &\mu_{F_i^9} = \exp\left[\frac{-(x-3)^2}{2}\right], \quad \mu_{F_i^{10}} = \exp\left[\frac{-(x-4)^2}{2}\right], \\ &\mu_{F_i^{11}} = \exp\left[\frac{-(x-5)^2}{2}\right]. \end{aligned}$$

The simulation is carried out under the zero initial condition for two cases. In the first case, $T_L = 1.5$ and in the second case,

$$T_L = \begin{cases} 1.5, & 0 \leq t \leq 1, \\ 3, & t \geq 1. \end{cases}$$

Figures 7.1 and 7.2 show the simulation results for the first case and Figs. 7.3 and 7.4 show the simulation results for the second case. From these figures, It is clear that the tracking performance has been achieved good results. This means that the proposed controller can track the reference signal satisfactorily even under parameter uncertainties and load torque disturbance.

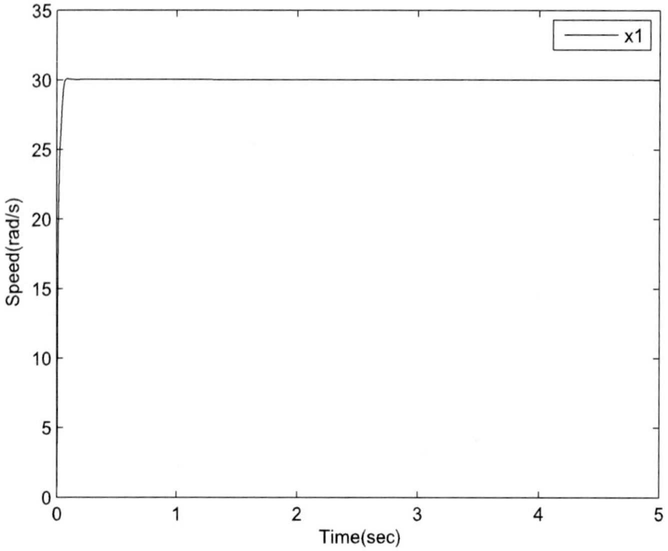

Fig. 7.1 The curve of the rotor speed x_1 in the first case

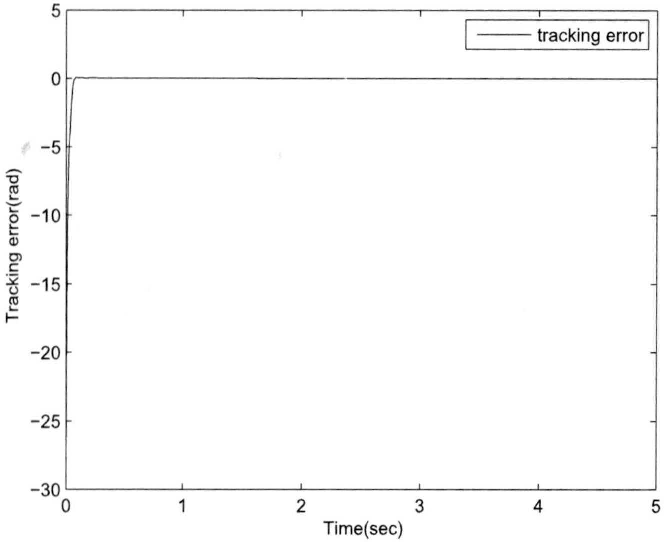

Fig. 7.2 The curve of tracking error z_1 in the first case

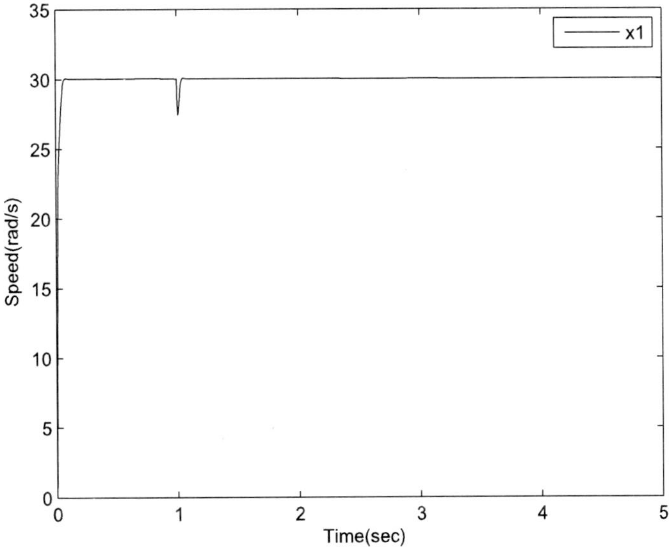

Fig. 7.3 The curve of the rotor speed x_1 in the second case

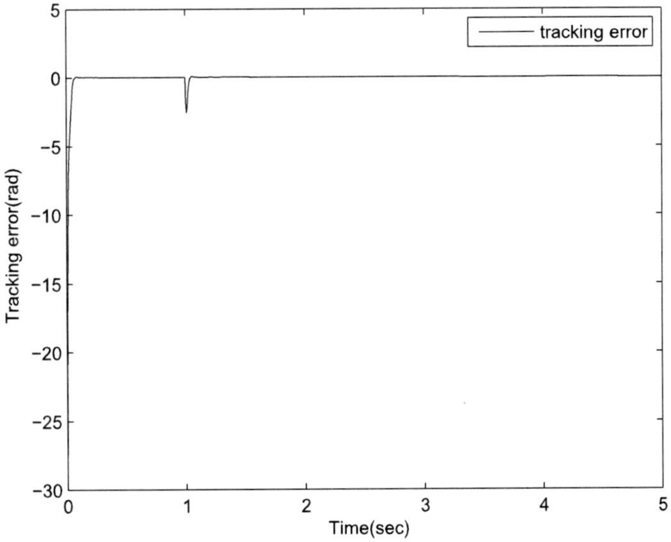

Fig. 7.4 The curve of tracking error z_1 in the second case

7.5 Conclusion

Based on the adaptive fuzzy control method and backstepping control technique, an adaptive fuzzy control scheme is proposed for PMSM. The proposed controllers ensure that the tracking error converges to a small neighborhood of the origin and all the closed-loop signals are bounded. Simulation results show the effectiveness of the proposed method.

References

1. Li, D., Zhang, X.H., Yang, D., Wang, S.L.: Fuzzy control of chaos in permanent magnet synchronous motor with parameter uncertainties. Acta Phys. Sin. **58**(3), 1432–1440 (2009)
2. Jung, J.W., Han, H.C., Lee, D.M.: Implementation of a robust fuzzy adaptive speed tracking control system for permanent magnet synchronous motors. J. Power Electron. **12**(6), 904–911 (2012)
3. Chen, B., Liu, X.P., Ge, S.S., Lin, C.: Adaptive fuzzy control of a class of nonlinear systems by fuzzy approximation approach. IEEE Trans. Fuzzy Syst. **20**(6), 1012–1021 (2012)
4. Ebrahimi, B.M., Faiz, J.: Diagnosis and performance analysis of threephase permanent magnet synchronous motors with static, dynamic and mixed eccentricity. IET Electric Power Appl. **4**(1), 53–66 (2010)
5. Tong, S.C., Li, H.H.: Fuzzy adaptive sliding model control for MIMO nonlinear systems. IEEE Trans. Fuzzy Syst. **11**(3), 354–360 (2003)
6. Lee, H., Tomizuka, M.: Robust adaptive control using a universal approximator for SISO nonlinear systems. IEEE Trans. Fuzzy Syst. **8**, 95–106 (2000)
7. Krstic, M., Kanellakopoulos, I., Kokotovic, P.: Nonlinear Adaptive Control and Design. Wiley, New York (1995)
8. Liu, X.P., Gu, G.X., Zhou, K.M.: Robust stabilization of MIMO nonlinear systems by backstepping. Automatica **35**(2), 987–992 (1999)
9. Hu, J., Zou, J.: Adaptive backstepping control of permanent magnet synchronous motors with parameter uncertainties. Control Decis. **21**(11), 1264–1269 (2006)
10. Zadeh, L.A.: Fuzzy sets. Inf. Technol. Control **8**(3), 338–353 (1965)
11. Cetin, E., Oguz, U., Hasan, H.S.: A neuro-fuzzy controller for speed control of a permanent magnet synchronous motor drive. Expert Syst. Appl. **34**, 657–664 (2006)
12. Tong, S.C., Li, H.H.: Direct adaptive fuzzy output tracking control of nonlinear systems. Fuzzy Sets Syst. **128**, 107–115 (2002)
13. Kung, Y.S., Tsai, M.H.: FPGA-based speed control IC for PMSM drive with adaptive fuzzy control. IEEE Trans. Power Electron. **22**(6), 2476–2486 (2007)
14. Wang, Y.H., Fan, Y.Q., Wang, Q.Y., Zhang, Y.: Adaptive fuzzy synchronization for a class of chaotic systems with unknown nonlinearities and disturbances. Nonlinear Dyn. **69**(3), 1167–1176 (2012)
15. Li, T., Chen, A.Q., Li, D.J.: Time-varying tan-type barrier Lyapunov function-based adaptive fuzzy control for switched systems with unknown dead zone. IEEE Access **7**, 110928–110935 (2019)
16. Pillay, P., Krishnan, R.: Modeling of permanent magnet motor drives. IEEE Trans. Ind. Electron. **35**(4), 537–541 (1998)
17. Wang, L.X., Mendel, J.M.: Fuzzy basis functions, universal approximation, and orthogonal least squares learning. IEEE Trans. Neural Netw. **3**(5), 807–814 (1992)

Chapter 8
Adaptive Fuzzy Backstepping Position Tracking Control for PMSM

The position tracking control problem of the permanent magnet synchronous motor (PMSM) with parameter uncertainties and load torque disturbance is addressed. Fuzzy logic systems are used to approximate nonlinearities and adaptive backstepping technique is employed to structure controllers. The proposed adaptive fuzzy controllers guarantee that the tracking error converges to a small neighborhood of the origin. Compared with the conventional backstepping method, the proposed fuzzy controller's structure is very simple and easy to be applied in practice. The simulation results illustrate the effectiveness of the proposed results.

8.1 Introduction

Modern electrical drives based on the PMSM are of great interest for industrial applications due to their high speed, high efficiency, high power density and large torque to inertia ratio [1–5]. However, the performance of the PMSM is very sensitive to external load disturbances and parameter variations in the plant because their dynamic model is usually multivariable, coupled and highly nonlinear [6–8]. In recent years, fuzzy logic control (FLC) [9, 10] has been found as one of the popular and conventional tools in functional approximations to deal with uncertain information [11–13]. It provides an effective way to design a control system as one of the important applications in the area of control engineering [14].

In this chapter, an adaptive fuzzy control approach is proposed for position tracking control of the PMSM drive system via backstepping [15, 16]. During the controller design process, fuzzy logic systems are used to approximate the nonlinearities, adaptive technique and backstepping are used to construct fuzzy controllers [17]. Moreover, the proposed controllers ensure that the tracking error converges to a small neighborhood of the origin and all the closed-loop signals are bounded.

J. Yu et al., *Intelligent Backstepping Control for the Alternating-Current Drive Systems*, Studies in Systems, Decision and Control 349,
https://doi.org/10.1007/978-3-030-67723-7_8

To verify the advantage of the proposed control method, a comparison between the classical backstepping controller and ours is studied [18]. The simulation results illustrate the effectiveness and robustness of the proposed controller [19, 20].

8.2 Mathematical Model of the PMSM Drive System and Preliminaries

In this section, some preparatory knowledge of a PMSM will be first introduced. To gain the mathematical model of a PMSM, the following assumptions are made.

Assumption 8.1 [21] Saturation and iron losses are ignored although it can be taken into account by parameter changes.

Assumption 8.2 [21] The back emf is sinusoidal.

From (1.16), the dynamic model of a PMSM motor can be described by the following differential equations:

$$
\begin{aligned}
\dot{x}_1 &= x_2,\\
\dot{x}_2 &= \frac{a_1}{J}x_3 + \frac{a_2}{J}x_3x_4 - \frac{B}{J}x_2 - \frac{T_L}{J},\\
\dot{x}_3 &= b_1x_3 + b_2x_2x_4 + b_3x_2 + b_4u_q,\\
\dot{x}_4 &= c_1x_4 + c_2x_2x_3 + c_3u_d,
\end{aligned}
\tag{8.1}
$$

where x_1 is the rotor position, x_2 denotes the rotor angular velocity, x_3 and x_4 stand for the $d-q$ axis currents, u_d and u_q are the $d-q$ axis voltages, J means the rotor moment of inertia, B is the viscous friction coefficient, T_L is the load torque.

8.3 Adaptive Fuzzy Controller with the Backstepping Technique

For the system (8.1), the backstepping design procedure includes 4 steps. At each design step, a virtual control function $\alpha_i\,(i = 1, 2)$ will be constructed using an appropriate Lyapunov function V. In the last step, a real controller is constructed to control the system.

Step 1: For the reference signal x_d, define the tracking error variable as $z_1 = x_1 - x_d$. From the first subsystem of (8.1), the error dynamic system is computed by $\dot{z}_1 = x_2 - \dot{x}_d$.

Choose Lyapunov function candidate as $V_1 = \frac{1}{2}z_1^2$, then the time derivative of V_1 is given by

$$
\dot{V}_1 = z_1\dot{z}_1 = z_1(x_2 - \dot{x}_d). \tag{8.2}
$$

Now, construct the virtual control law α_1 as

$$\alpha_1(x_1, x_d, \dot{x}_d) = -k_1 z_1 + \dot{x}_d, \tag{8.3}$$

with $k_1 > 0$ being a design parameter and defining $z_2 = x_2 - \alpha_1$. By using (8.3), (8.2) can be rewritten as following

$$\dot{V}_1 = -k_1 z_1^2 + z_1 z_2.$$

Step 2: Differentiating z_2 gives

$$\dot{z}_2 = \frac{a_1}{J} x_3 + \frac{a_2}{J} x_3 x_4 - \frac{B}{J} x_2 - \frac{T_L}{J} - \dot{\alpha}_1. \tag{8.4}$$

Now, choose the Lyapunov function candidate as $V_2 = V_1 + \frac{J}{2} z_2^2$. Obviously, time derivative of V_2 is written as

$$\dot{V}_2 = \dot{V}_1 + J z_2 \dot{z}_2 = -k_1 z_1^2 + z_2(a_1 x_3 + z_1 + a_2 x_3 x_4 - B x_2 - T_L - J\dot{\alpha}_1). \tag{8.5}$$

Then the virtual control α_2 is constructed as

$$\alpha_2(x_1, x_2, x_d, \dot{x}_d, \ddot{x}_d, \hat{B}, \hat{T}_L, \hat{J}) = \frac{1}{a_1}(-k_2 z_2 - z_1 + \hat{B} x_2 + \hat{T}_L + \hat{J}\dot{\alpha}_1), \tag{8.6}$$

where $\hat{B}$, $\hat{T}_L$ and $\hat{J}$ are their estimations of B, T_L and J, respectively. Adding and subtracting α_2 in the bracket in (8.5) show that

$$\dot{V}_2 = -k_1 z_1^2 - k_2 z_2^2 + a_1 z_2 z_3 + a_2 z_2 x_3 x_4 + z_2(\hat{B} - B)x_2 + z_2(\hat{T}_L - T_L) + z_2(\hat{J} - J)\dot{\alpha}_1, \tag{8.7}$$

with $k_2 > 0$ being a positive design parameter and $z_3 = x_3 - \alpha_2$.

Remark 8.1 In the realistic model of the PMSM, the system parameters B, T_L and J may be unknown, they cannot be used to construct the control signal unless we specify their corresponding adaptation law. Thus, in this paper, the adaptive technique is used to estimate these parameters on-line. ♦

Step 3: Differentiating z_3 results in the following differential equation

$$\dot{z}_3 = \dot{x}_3 - \dot{\alpha}_2 = b_1 x_3 + b_2 x_2 x_4 + b_3 x_2 + b_4 u_q - \dot{\alpha}_2.$$

Choose the following Lyapunov function candidate as $V_3 = V_2 + \frac{1}{2} z_3^2$. Furthermore, differentiating V_3 yields

$$\begin{aligned}\dot{V}_3 &= \dot{V}_2 + z_3\dot{z}_3 = \dot{V}_2 + z_3(b_1x_3 + b_2x_2x_4 + b_3x_2 + b_4u_q - \dot{\alpha}_2)\\ &= -k_1z_1^2 - k_2z_2^2 + a_2z_2x_3x_4 + z_2(\hat{B} - B)x_2 + z_2(\hat{T}_L - T_L)\\ &\quad + z_2(\hat{J} - J)\dot{\alpha}_1 + z_3(f_3 + b_4u_q),\end{aligned} \tag{8.8}$$

where

$$\begin{aligned}\dot{\alpha}_1 &= \dot{x}_1 - \dot{x}_d = x_2 - \dot{x}_d,\\ \dot{\alpha}_2 &= \sum_{i=1}^{2}\frac{\partial\alpha_2}{\partial x_i}\dot{x}_i + \sum_{i=0}^{2}\frac{\partial\alpha_2}{\partial x_d^{(i)}}x_d^{(i+1)} + \frac{\partial\alpha_2}{\partial\hat{B}}\dot{\hat{B}} + \frac{\partial\alpha_2}{\partial\hat{T}_L}\dot{\hat{T}}_L + \frac{\partial\alpha_2}{\partial\hat{J}}\dot{\hat{J}}\\ &= \frac{\partial\alpha_2}{\partial x_1}x_2 + \frac{\partial\alpha_2}{\partial x_2}\left(\frac{a_1}{J}x_3 + \frac{a_2}{J}x_3x_4 - \frac{B}{J}x_2 - \frac{T_L}{J}\right)\\ &\quad + \sum_{i=0}^{2}\frac{\partial\alpha_2}{\partial x_d^{(i)}}x_d^{(i+1)} + \frac{\partial\alpha_2}{\partial\hat{B}}\dot{\hat{B}} + \frac{\partial\alpha_2}{\partial\hat{T}_L}\dot{\hat{T}}_L + \frac{\partial\alpha_2}{\partial\hat{J}}\dot{\hat{J}},\\ f_3(Z_3) &= b_1x_3 + b_2x_2x_4 + b_3x_2 + a_1z_2 - \dot{\alpha}_2,\\ Z_3 &= [x_1, x_2, x_3, x_4, x_d, \dot{x}_d, \ddot{x}_d, \hat{B}, \hat{T}_L, \hat{J}]^T.\end{aligned} \tag{8.9}$$

Notice that f_3 contains the derivative of α_2, the unknown system parameters B, T_L and J appear in the expression of f_3. This will make the classical adaptive backstepping design become very complex and troubled, and the designed control law u_q will have a complex structure. To avoid this trouble and simplify the control signal structure, we will employ the fuzzy logic system to approximate the nonlinear function f_3. As shown later, the design procedure of u_q becomes simple and u_q has a simple structure. For any given $\varepsilon_3 > 0$, there exists a fuzzy logic system $W_3^TS_3(Z_3)$ such that

$$f_3(Z_3) = W_3^TS_3(Z_3) + \delta_3(Z_3), \tag{8.10}$$

with $\delta_3(Z_3)$ being the approximation error and satisfying $|\delta_3| \le \varepsilon_3$. Consequently, a simple computing method produces the following inequality.

$$z_3f_3(Z_3) = z_3\left(W_3^TS_3 + \delta_3\right) \le \frac{1}{2l_3^2}z_3^2\|W_3\|^2S_3^2 + \frac{1}{2}l_3^2 + \frac{1}{2}z_3^2 + \frac{1}{2}\varepsilon_3^2. \tag{8.11}$$

Thus, it follows immediately from substituting (8.11) into (8.8) that

$$\begin{aligned}\dot{V}_3 &\le -k_1z_1^2 - k_2z_2^2 + a_2z_2x_3x_4 + z_2(\hat{B} - B)x_2 + z_2(\hat{T}_L - T_L)\\ &\quad + z_2(\hat{J} - J)\dot{\alpha}_1 + \frac{1}{2l_3^2}z_3^2\|W_3\|^2S_3^2 + \frac{1}{2}l_3^2 + \frac{1}{2}z_3^2 + \frac{1}{2}\varepsilon_3^2 + z_3b_4u_q.\end{aligned}$$

At this present stage, the control input u_q is designed as

$$u_q = \frac{1}{b_4}(-k_3 z_3 - \frac{1}{2} z_3 - \frac{1}{2l_3^2} z_3 \hat{\theta} S_3^2), \tag{8.12}$$

where $\hat{\theta}$ is the estimation of the unknown constant θ which will be specified later. Furthermore using the equality (8.12), it can be verified easily that

$$\dot{V}_3 \leq -\sum_{i=1}^{3} k_i z_i^2 + a_2 z_2 x_3 x_4 + z_2(\hat{B} - B)x_2 + z_2(\hat{T}_L - T_L)$$
$$+ z_2(\hat{J} - J)\dot{\alpha}_1 + \frac{1}{2l_3^2} z_3^2(\|W_3\|^2 - \hat{\theta})S_3^2 + \frac{1}{2}l_3^2 + \frac{1}{2}\varepsilon_3^2.$$

Step 4: At this step, we will construct the control law u_d. To this end, define $z_4 = x_4$ and choose the following Lyapunov function candidate as $V_4 = V_3 + \frac{1}{2}z_4^2$. Then the derivative of V_4 is written as

$$\begin{aligned}
\dot{V}_4 &= \dot{V}_3 + z_4 \dot{z}_4, \\
&\leq -\sum_{i=1}^{3} k_i z_i^2 + \frac{1}{2l_3^2} z_3^2(\|W_3\|^2 - \hat{\theta})S_3^2 + \frac{1}{2}l_3^2 + \frac{1}{2}\varepsilon_3^2 + z_2(\hat{B} - B)x_2 \\
&\quad + z_2(\hat{T}_L - T_L) + z_2(\hat{J} - J)\dot{\alpha}_1 + z_4\left(f_4(Z_4) + c_3 u_d\right),
\end{aligned} \tag{8.13}$$

where $f_4(Z_4) = a_2 z_2 x_3 + c_1 z_4 + c_2 x_2 x_3$. Similarly, the fuzzy logic system $W_4^T S_4(Z_4)$ is utilized to approximate the nonlinear function f_4 such that for given $\varepsilon_4 > 0$,

$$z_4 f_4(Z_4) \leq \frac{1}{2l_4^2} z_4^2 \|W_4\|^2 S_4^2 + \frac{1}{2}l_4^2 + \frac{1}{2}z_4^2 + \frac{1}{2}\varepsilon_4^2. \tag{8.14}$$

Substituting (8.14) into (8.13) gives

$$\begin{aligned}
\dot{V}_4 &= \dot{V}_3 + z_4 \dot{z}_4, \\
&\leq -\sum_{i=1}^{3} k_i z_i^2 + \frac{1}{2l_3^2} z_3^2(\|W_3\|^2 - \hat{\theta})S_3^2 + \sum_{i=3}^{4} \frac{1}{2}(l_i^2 + \varepsilon_i^2) \\
&\quad + z_2(\hat{B} - B)x_2 + z_2(\hat{T}_L - T_L) + z_2(\hat{J} - J)\dot{\alpha}_1 \\
&\quad + \frac{1}{2l_4^2} z_4^2 \|W_4\|^2 S_4^2 + \frac{1}{2}z_4^2 + c_3 z_4 u_d.
\end{aligned} \tag{8.15}$$

Now design u_d as

$$u_d = \frac{-1}{c_3}(k_4 z_4 + \frac{1}{2} z_4 + \frac{1}{2l_4^2} z_4 \hat{\theta} S_4^2), \tag{8.16}$$

and define $\theta = \max\{\|W_3\|^2, \|W_4\|^2\}$. Furthermore, combining (8.15) with (8.16) results in

$$
\begin{aligned}
\dot{V}_4 \leq & -\sum_{i=1}^{4} k_i z_i^2 + \sum_{i=3}^{4} \frac{1}{2}(l_i^2 + \varepsilon_i^2) + z_2(\hat{B} - B)x_2 + z_2(\hat{T}_L - T_L) \\
& + z_2(\hat{J} - J)\dot{\alpha}_1 + \sum_{i=3}^{4} \frac{1}{2l_i^2} z_i^2 \left(\|W_i\|^2 - \hat{\theta}\right) S_i(Z)^T S_i(Z), \\
\leq & -\sum_{i=1}^{4} k_i z_i^2 + \sum_{i=3}^{4} \frac{1}{2}(l_i^2 + \varepsilon_i^2) + z_2(\hat{B} - B)x_2 + z_2(\hat{T}_L - T_L) \\
& + z_2(\hat{J} - J)\dot{\alpha}_1 + \sum_{i=3}^{4} \frac{1}{2l_i^2} z_i^2 S_i^T S_i \left(\theta - \hat{\theta}\right).
\end{aligned}
\tag{8.17}
$$

Introduce $\tilde{B}$, $\tilde{T}_L$, $\tilde{J}$ and $\tilde{\theta}$ as

$$
\begin{aligned}
\tilde{T}_L &= \hat{T}_L - T_L, \\
\tilde{B} &= \hat{B} - B, \\
\tilde{J} &= \hat{J} - J, \\
\tilde{\theta} &= \hat{\theta} - \theta,
\end{aligned}
\tag{8.18}
$$

and choose the following Lyapunov function candidate:

$$
V = V_4 + \frac{1}{2r_1}\tilde{T}_L^2 + \frac{1}{2r_2}\tilde{B}^2 + \frac{1}{2r_3}\tilde{J}^2 + \frac{1}{2r_4}\tilde{\theta}^2,
\tag{8.19}
$$

where $r_i, i = 1, 2, 3, 4$ are positive constant. By differentiating V and taking (8.17)–(8.19) into account, one has

$$
\begin{aligned}
\dot{V} \leq & -\sum_{i=1}^{4} k_i z_i^2 + \sum_{i=3}^{4} \frac{1}{2}(l_i^2 + \varepsilon_i^2) + z_2(\hat{B} - B)x_2 + z_2(\hat{T}_L - T_L) \\
& + z_2(\hat{J} - J)\dot{\alpha}_1 + \sum_{i=3}^{4} \frac{1}{2l_i^2} z_i^2 S_i^T S_i \left(\theta - \hat{\theta}\right) \\
& + \frac{1}{r_1}\tilde{T}_L\dot{\hat{T}}_L + \frac{1}{r_2}\tilde{B}\dot{\hat{B}} + \frac{1}{r_3}\tilde{J}\dot{\hat{J}} + \frac{1}{r_4}\tilde{\theta}\dot{\hat{\theta}}, \\
= & -\sum_{i=1}^{4} k_i z_i^2 + \sum_{i=3}^{4} \frac{1}{2}(l_i^2 + \varepsilon_i^2) + \frac{1}{r_1}\tilde{T}_L\left(r_1 z_2 + \dot{\hat{T}}_L\right) +
\end{aligned}
$$

$$\frac{1}{r_2}\tilde{B}\left(r_2 z_2 x_2 + \dot{\hat{B}}\right) + \frac{1}{r_3}\tilde{J}\left(r_3 z_2 \dot{\alpha}_1 + \dot{\hat{J}}\right)$$
$$+\frac{1}{r_4}\tilde{\theta}\left[-\sum_{i=3}^{4}\frac{r_4}{2l_i^2}z_i^2 S_i^T S_i + \dot{\hat{\theta}}\right]. \quad (8.20)$$

According to (8.20), the corresponding adaptive laws are chosen as follows:

$$\begin{aligned}
\dot{\hat{T}}_L &= -r_1 z_2 - m_1 \hat{T}_L, \\
\dot{\hat{B}} &= -r_2 z_2 x_2 - m_2 \hat{B}, \\
\dot{\hat{J}} &= -r_3 z_2 \dot{\alpha}_1 - m_3 \hat{J}, \\
\dot{\hat{\theta}} &= \sum_{i=3}^{4}\frac{r_4}{2l_i^2} z_i^2 S_i^T S_i - m_4 \hat{\theta},
\end{aligned} \quad (8.21)$$

where m_i for $i = 1, 2, 3, 4$ and l_i for $i = 3, 4$ are positive constant.

Remark 8.2 According to the proposed adaptive fuzzy controllers, it is clearly seen that the proposed controllers have a simpler structure. This means that the proposed adaptive fuzzy controllers are easy to be used in practical engineering. ♦

Theorem 8.3 *Consider the system (8.1) satisfying assumptions Figs. 8.1, 8.2 and the given reference signal x_d. Then under the action of the adaptive fuzzy controller (8.12) and (8.16), the tracking error of the closed-loop controlled system will converge into a sufficiently small neighborhood of the origin and all the closed-loop signals are bounded.*

Proof To address the stability analysis of the resulting closed-loop system, substitute (8.21) into (8.20) to obtain that

$$\dot{V} \leq -\sum_{i=1}^{4} k_i z_i^2 + \sum_{i=3}^{4}\frac{1}{2}(l_i^2 + \varepsilon_i^2) - \frac{m_1}{r_1}\tilde{T}_L\hat{T} - \frac{m_2}{r_2}\tilde{B}\hat{B} - \frac{m_3}{r_3}\tilde{J}\hat{J} - \frac{m_4}{r_4}\tilde{\theta}\hat{\theta}. \quad (8.22)$$

For the term $-\tilde{T}_L\hat{T}$, one has

$$-\tilde{T}_L\hat{T}_L = -\tilde{T}_L(\tilde{T}_L + T_L) \leq -\frac{1}{2}\tilde{T}_L^2 + \frac{1}{2}T_L^2.$$

Similarly, we have

$$\begin{aligned}
-\tilde{B}\hat{B} &\leq -\frac{1}{2}\tilde{B}^2 + \frac{1}{2}B^2, \\
-\tilde{J}\hat{J} &\leq -\frac{1}{2}\tilde{J}^2 + \frac{1}{2}J^2, \\
-\tilde{\theta}\hat{\theta} &\leq -\frac{1}{2}\tilde{\theta}^2 + \frac{1}{2}\theta^2.
\end{aligned}$$

Consequently, by using these inequalities (8.22) can be rewritten in the following form

$$
\begin{aligned}
\dot{V} &\leq -\sum_{i=1}^{4} k_i z_i^2 - \frac{m_1}{2r_1}\tilde{T}_L^2 - \frac{m_2}{2r_2}\tilde{B}^2 - \frac{m_3}{2r_3}\tilde{J}^2 - \frac{m_4}{2r_4}\tilde{\theta}^2 \\
&\quad + \sum_{i=3}^{4} \frac{1}{2}(l_i^2 + \varepsilon_i^2) + \frac{m_1}{2r_1}T_L^2 + \frac{m_2}{2r_2}B^2 + \frac{m_3}{2r_3}J^2 + \frac{m_4}{2r_4}\theta^2, \\
&\leq -a_0 V + b_0,
\end{aligned}
\tag{8.23}
$$

where $a_0 = \min\left\{2k_1, \frac{2k_2}{J}, 2k_3, 2k_4, m_1, m_2, m_3, m_4\right\}$ and $b_0 = \sum_{i=3}^{4} \frac{1}{2}(l_i^2 + \varepsilon_i^2) + \frac{m_1}{2r_1} T_L^2 + \frac{m_2}{2r_2}B^2 + \frac{m_3}{2r_3}J^2 + \frac{m_4}{2r_4}\theta^2$. Furthermore, (8.23) implies that

$$
V(t) \leq (V(t_0) - \frac{b_0}{a_0})e^{-a_0(t-t_0)} + \frac{b_0}{a_0} \leq V(t_0) + \frac{b_0}{a_0}, \forall t \geqslant t_0. \tag{8.24}
$$

As a result, all $z_i (i = 1, 2, 3, 4)$, $\tilde{T}_L$, $\tilde{B}$, $\tilde{J}$ and $\tilde{\theta}$ belong to the compact set

$$
\Omega = \left\{ (z_i, \tilde{T}_L, \tilde{B}, \tilde{J}, \tilde{\theta}) | V \leq V(t_0) + \frac{b_0}{a_0}, \forall t \geqslant t_0 \right\}.
$$

Namely, all the signals in the closed-loop system are bounded. Especially, from (8.24) we have

$$
\lim_{t \to \infty} z_1^2 \leq \frac{2b_0}{a_0}.
$$

From the definitions of a_0 and b_0, it is clear that to get a small tracking error we can take r_i large and l_i and ε_i small enough after giving the parameters k_i and m_i. ■

8.4 A Comparison with the Conventional Backstepping Design

8.4.1 Conventional Backstepping Design

The control of the PMSM based on conventional backstepping technique is reviewed here.

Step 1: For the reference signal x_d, define the tracking error variable as $z_1 = x_1 - x_d$. From the first subsystem of (8.1), the error dynamic system is given by $\dot{z}_1 = x_2 - \dot{x}_d$.

Choose Lyapunov function candidate as $V_1 = \frac{1}{2}z_1^2$, then the time derivative of V_1 is given by

$$\dot{V}_1 = z_1\dot{z}_1 = z_1(x_2 - \dot{x}_d). \tag{8.25}$$

Now, construct the virtual control law α_1 as

$$\alpha_1(x_1, x_d, \dot{x}_d) = -k_1 z_1 + \dot{x}_d, \tag{8.26}$$

where $k_1 > 0$ is a design parameter. Defining $z_2 = x_2 - \alpha_1$ and substituting (8.26) into (8.25) yield

$$\dot{V}_1 = -k_1 z_1^2 + z_1 z_2.$$

Step 2: Differentiating z_2 gives

$$\dot{z}_2 = \frac{a_1}{J}x_3 + \frac{a_2}{J}x_3x_4 - \frac{B}{J}x_2 - \frac{T_L}{J} - \dot{\alpha}_1. \tag{8.27}$$

Now, choose the Lyapunov function candidate as $V_2 = V_1 + \frac{J}{2}z_2^2$. Obviously, the time derivative of V_2 is given by

$$\dot{V}_2 = \dot{V}_1 + Jz_2\dot{z}_2 = -k_1 z_1^2 + z_2(a_1x_3 + z_1 + a_2x_3x_4 - Bx_2 - T_L - J\dot{\alpha}_1). \tag{8.28}$$

Since the parameters B, T_L and J are unknown, they cannot be used to construct the control signal. Thus, let $\hat{B}$, $\hat{T}_L$ and $\hat{J}$ be their estimations of B, T_L and J, respectively. The virtual control α_2 is constructed as

$$\alpha_2 = \frac{1}{a_1}(-k_2z_2 - z_1 + \hat{B}x_2 + \hat{T}_L + \hat{J}\dot{\alpha}_1), \tag{8.29}$$

where $k_2 > 0$ is a positive design parameter. Adding and subtracting α_2 in the bracket in (8.28) show that

$$\dot{V}_2 = -k_1z_1^2 - k_2z_2^2 + a_1z_2z_3 + a_2z_2x_3x_4 + z_2(\hat{B} - B)x_2 + z_2(\hat{T}_L - T_L) + z_2(\hat{J} - J)\dot{\alpha}_1, \tag{8.30}$$

with $z_3 = x_3 - \alpha_2$.

Step 3: Differentiating z_3 results in the following differential equation.

$$\dot{z}_3 = \dot{x}_3 - \dot{\alpha}_2 = b_1x_3 + b_2x_2x_4 + b_3x_2 + b_4u_q - \dot{\alpha}_2.$$

Choose the following Lyapunov function candidate as $V_3 = V_2 + \frac{1}{2}z_3^2$. Furthermore, differentiating V_3 yields

$$\begin{aligned}\dot{V}_3 &= \dot{V}_2 + z_3\dot{z}_3 = \dot{V}_2 + z_3(b_1x_3 + b_2x_2x_4 + b_3x_2 + b_4u_q - \dot{\alpha}_2),\\ &= -k_1z_1^2 - k_2z_2^2 + a_2z_2x_3x_4 + b_2x_2z_3x_4 + z_2(\hat{B} - B)x_2 + z_2(\hat{T}_L - T_L)\\ &\quad + z_2(\hat{J} - J)\dot{\alpha}_1 + z_3(b_1x_3 + b_3x_2 + a_1z_2 - \dot{\alpha}_2 + b_4u_q),\end{aligned} \tag{8.31}$$

where

$$\begin{aligned}\dot{\alpha}_1 &= \dot{x}_1 - \dot{x}_d = x_2 - \dot{x}_d,\\ \dot{\alpha}_2 &= \sum_{i=1}^{2}\frac{\partial\alpha_2}{\partial x_i}\dot{x}_i + \sum_{i=0}^{2}\frac{\partial\alpha_2}{\partial x_d^{(i)}}x_d^{(i+1)} + \frac{\partial\alpha_2}{\partial\hat{B}}\dot{\hat{B}} + \frac{\partial\alpha_2}{\partial\hat{T}_L}\dot{\hat{T}}_L + \frac{\partial\alpha_2}{\partial\hat{J}}\dot{\hat{J}},\\ &= \frac{\partial\alpha_2}{\partial x_1}x_2 + \frac{\partial\alpha_2}{\partial x_2}\left(\frac{a_1}{J}x_3 + \frac{a_2}{J}x_3x_4 - \frac{B}{J}x_2 - \frac{T_L}{J}\right)\\ &\quad + \sum_{i=0}^{2}\frac{\partial\alpha_2}{\partial x_d^{(i)}}x_d^{(i+1)} + \frac{\partial\alpha_2}{\partial\hat{B}}\dot{\hat{B}} + \frac{\partial\alpha_2}{\partial\hat{T}_L}\dot{\hat{T}}_L + \frac{\partial\alpha_2}{\partial\hat{J}}\dot{\hat{J}}.\end{aligned}$$

Then the control input u_q is designed as

$$\begin{aligned}u_q &= \frac{1}{b_4}(-k_3z_3 - b_1x_3 - b_3x_2 - a_1z_2 + \dot{\alpha}_2),\\ &= \frac{1}{b_4}(-k_3z_3 - b_1x_3 - b_3x_2 - a_1z_2 + \frac{\partial\alpha_2}{\partial x_1}x_2 + \frac{\partial\alpha_2}{\partial x_2}\left(\frac{a_1}{J}x_3 + \frac{a_2}{J}x_3x_4 - \frac{B}{J}x_2 - \frac{T_L}{J}\right)\\ &\quad + \sum_{i=0}^{2}\frac{\partial\alpha_2}{\partial x_d^{(i)}}x_d^{(i+1)} + \frac{\partial\alpha_2}{\partial\hat{B}}\dot{\hat{B}} + \frac{\partial\alpha_2}{\partial\hat{T}_L}\dot{\hat{T}}_L + \frac{\partial\alpha_2}{\partial\hat{J}}\dot{\hat{J}}),\end{aligned} \tag{8.32}$$

with $k_3 > 0$. Using the equality (8.32), the derivative of V_3 becomes as

$$\dot{V}_3 \le -\sum_{i=1}^{3}k_iz_i^2 + a_2z_2x_3x_4 + b_2x_2z_3x_4 + z_2(\hat{B} - B)x_2 + z_2(\hat{T}_L - T_L) + z_2(\hat{J} - J)\dot{\alpha}_1.$$

Step 4: At this step, we will construct the control law u_d. To this end, define $z_4 = x_4$ and choose the following Lyapunov function candidate as $V_4 = V_3 + \frac{1}{2}z_4^2$. Then the derivative of V_4 is given by

$$\begin{aligned}\dot{V}_4 &= \dot{V}_3 + z_4\dot{z}_4,\\ &\le -\sum_{i=1}^{3}k_iz_i^2 + z_2(\hat{B} - B)x_2 + z_2(\hat{T}_L - T_L) + z_2(\hat{J} - J)\dot{\alpha}_1\\ &\quad + z_4\left(a_2z_2x_3 + b_2x_2z_3 + c_1z_4 + c_2x_2x_3 + c_3u_d\right).\end{aligned} \tag{8.33}$$

Then the control input u_d is design as

$$u_d = -\frac{1}{c_3}(k_4 z_4 + a_2 z_2 x_3 + b_2 x_2 z_3 + c_1 z_4 + c_2 x_2 x_3), \tag{8.34}$$

with $k_4 > 0$.

Remark 8.4 So far, by comparing the adaptive fuzzy controllers Eqs. (8.12) and (8.16) with the conventional backstepping controllers Eqs. (8.32) and (8.34), it can be seen clearly that the expression of backstepping controllers (8.32) and (8.34) would be much more complicated than that of the new controllers (8.12) and (8.16). The number of terms in the expression of (8.32) and (8.34) is much larger. This drawback is called the "explosion of terms ". ♦

8.4.2 *Simulation*

To give a further comparison, the proposed adaptive fuzzy controller (8.12) and (8.16) and the classical backstepping controllers (8.32) and (8.34) will be used to control the following real system, respectively. The simulation is run for PMSM with the parameters:

$$J = 0.00379\,\mathrm{Kgm}^2, \quad \mathrm{R_s} = 0.68, \quad \mathrm{L_d} = 0.00315\mathrm{H},$$
$$L_q = 0.00285\mathrm{H}, \quad \mathrm{B} = 0.001158\,\mathrm{Nm/(rad/s)}, \quad \mathrm{fl} = 0.1245\mathrm{H}, \quad \mathrm{n_p} = 3,$$

The simulation is carried out under the zero initial condition. The reference signal was taken as $x_d = \sin(t)$ and

$$T_L = \begin{cases} 1.5, & 0 \le t \le 5, \\ 3, & t \ge 5. \end{cases}$$

Now, the proposed adaptive fuzzy controllers are used to control this PMSM motor. The control parameters are chosen as follows:

$$k_1 = 75, \quad k_2 = 25, \quad k_3 = 35, \quad k_4 = 40, \quad r_1 = r_2 = r_3 = r_4 = 0.25,$$
$$m_1 = m_2 = m_3 = m_4 = 0.005, \quad l_3 = l_4 = 0.5.$$

The fuzzy membership functions are chosen as:

$$\mu_{F_i^1} = \exp\left[\frac{-(x+5)^2}{2}\right], \quad \mu_{F_i^2} = \exp\left[\frac{-(x+4)^2}{2}\right],$$
$$\mu_{F_i^3} = \exp\left[\frac{-(x+3)^2}{2}\right], \quad \mu_{F_i^4} = \exp\left[\frac{-(x+2)^2}{2}\right],$$
$$\mu_{F_i^5} = \exp\left[\frac{-(x+1)^2}{2}\right], \quad \mu_{F_i^6} = \exp\left[\frac{-(x-0)^2}{2}\right],$$

$$\mu_{F_i^7} = \exp\left[\frac{-(x-1)^2}{2}\right], \quad \mu_{F_i^8} = \exp\left[\frac{-(x-2)^2}{2}\right],$$
$$\mu_{F_i^9} = \exp\left[\frac{-(x-3)^2}{2}\right], \quad \mu_{F_i^{10}} = \exp\left[\frac{-(x-4)^2}{2}\right],$$
$$\mu_{F_i^{11}} = \exp\left[\frac{-(x-5)^2}{2}\right].$$

Next, the classical backstepping controllers (8.32) and (8.34) given in conventional backstepping design is also utilized to control the systems. The corresponding controller parameters are taken as

$$k_1 = 40, \quad k_2 = 25, \quad k_3 = 25, \quad k_4 = 25.$$

The simulation results for both cases of adaptive fuzzy control and classical backstepping control are shown by Figs. 8.1, 8.2, 8.3, 8.4, 8.5, 8.6, 8.7 and 8.8. Figures 8.1, 8.2, 8.3 and 8.4 display the system outputs and the reference signals for both control approaches, and Figs. 8.5, 8.6, 8.7 and 8.8 show the control input signals. From Figs. 8.1, 8.2, 8.3 and 8.4, it is seen clearly that under the actions of controllers (8.12) and (8.16) and the controllers (8.32) and (8.34), the system outputs follow the desired reference signals well. Figures 8.9 and 8.10 display the curves of i_d and i_q. Form the simulation, it is seen clearly that the proposed controllers can trace the reference signal quite well. So far, by comparing the above two set different controllers, it is easy to see that the proposed adaptive fuzzy controllers have a much more simple structure than the classical ones. This means that the proposed adaptive fuzzy controllers are easy to be implemented in practical engineering.

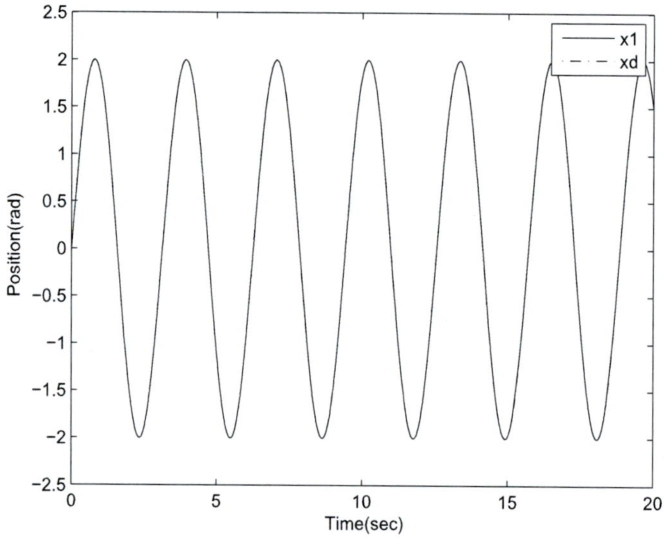

Fig. 8.1 Trajectories of the x_1 and x_{1d} for adaptive fuzzy control

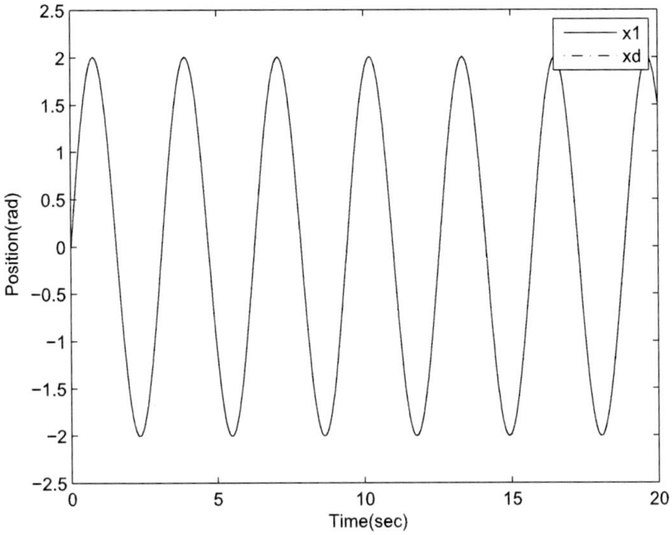

Fig. 8.2 Trajectories of the x_1 and x_{1d} for classical backstepping

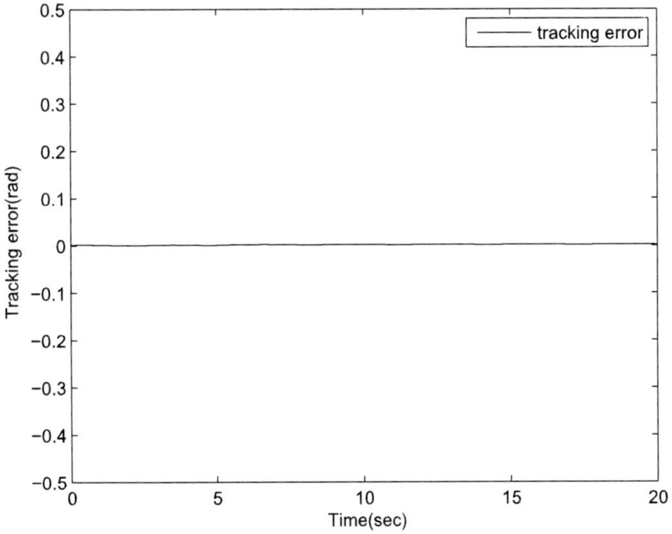

Fig. 8.3 Tracking error between the x_4 and x_{4d} for adaptive fuzzy control

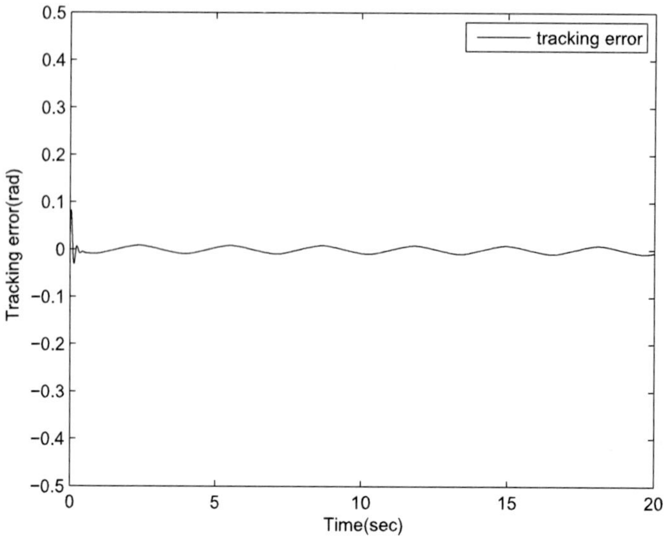

Fig. 8.4 Tracking error between the x_4 and x_{4d} for classical backstepping

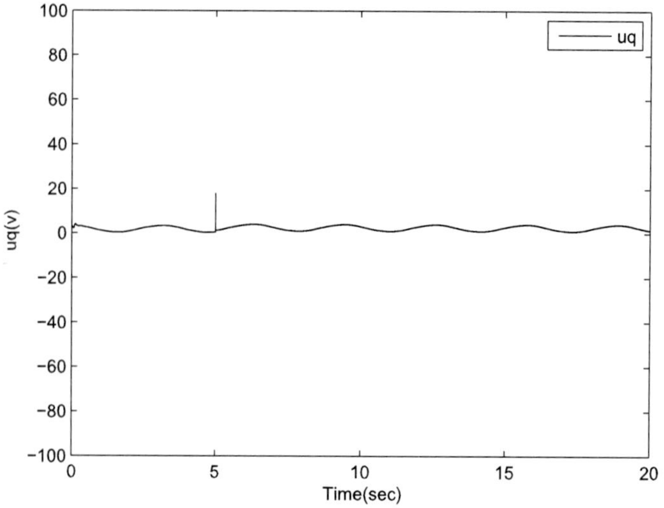

Fig. 8.5 Curve of the u_q for adaptive fuzzy control

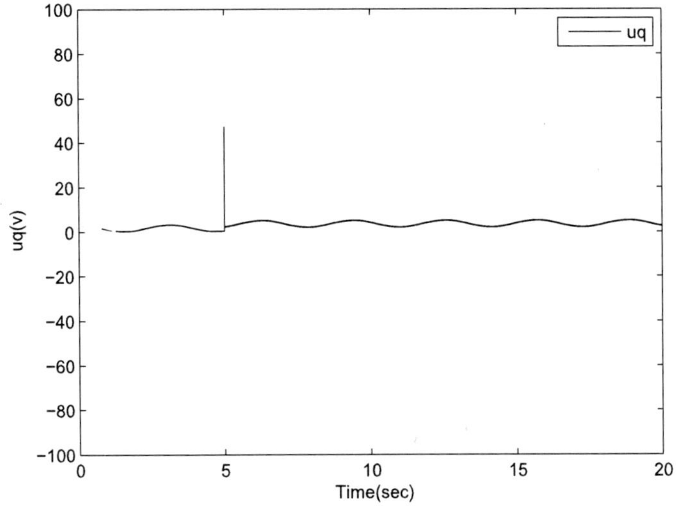

Fig. 8.6 Curve of the u_q for classical backstepping

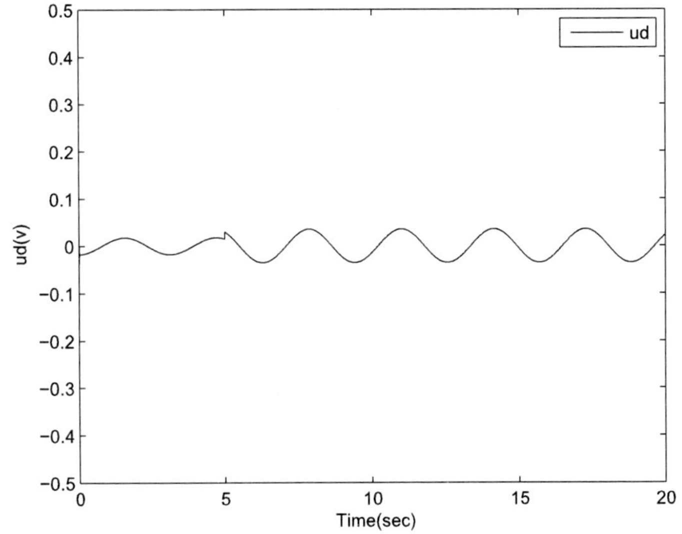

Fig. 8.7 Curve of the u_d for adaptive fuzzy control

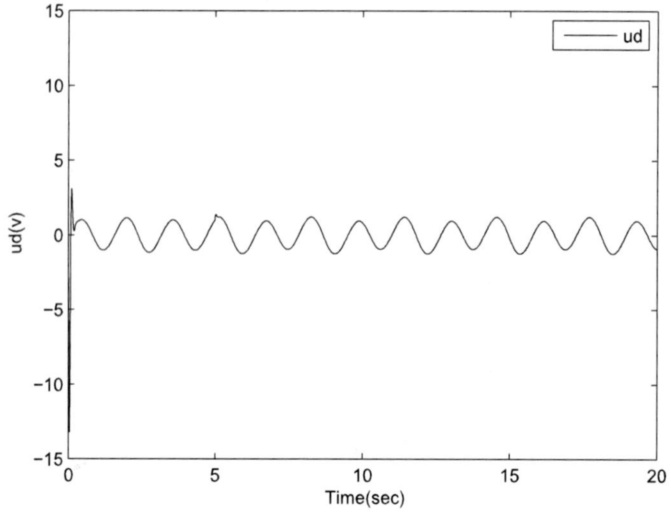

Fig. 8.8 Curve of the u_d for classical backstepping

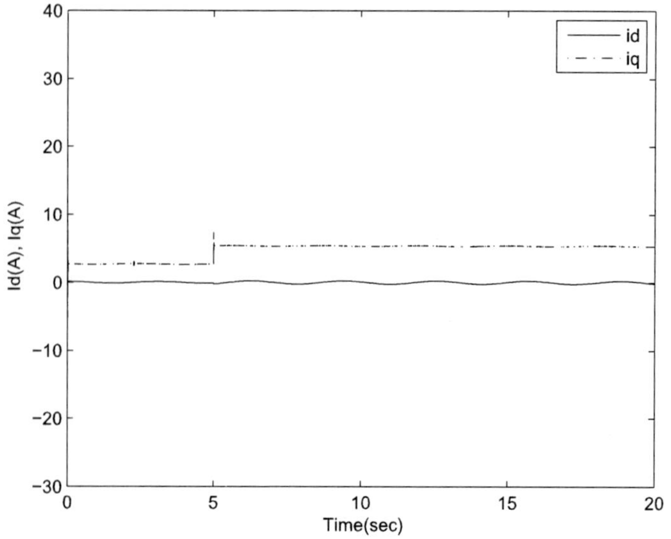

Fig. 8.9 Curves of the i_d, i_q for adaptive fuzzy control

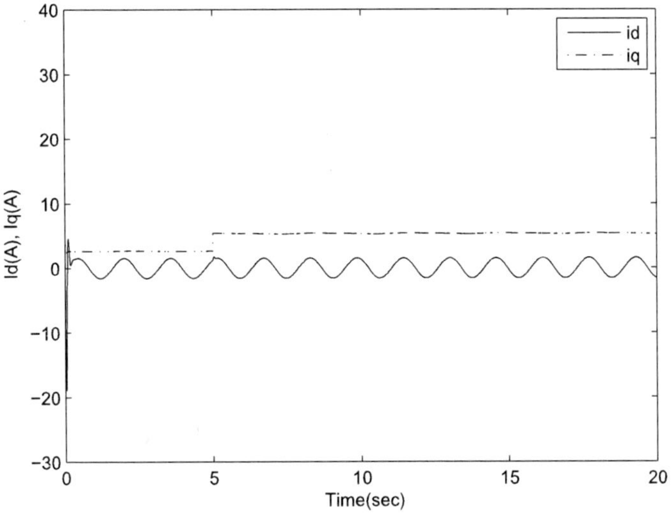

Fig. 8.10 Curves of the i_d, i_q for classical backstepping

8.5 Conclusion

Based on the adaptive fuzzy control approach and backstepping technique, an adaptive fuzzy control scheme is proposed to control a permanent magnet synchronous motor. The proposed controllers guarantee that the tracking error converges to a small neighborhood of the origin and all the closed-loop signals are bounded. Simulation results illustrate the effectiveness of the presented method.

References

1. Hu, J.H., Zou, J.B.: Adaptive backstepping control of permanent magnet synchronous motors with parameter uncertainties. Control Decis. **21**(11), 1264–9 (2006)
2. Liu, X.P., Gu, G.X., Zhou, K.M.: Robust stabilization of MIMO nonlinear systems by backstepping. Automatica **35**(5), 987–992 (1999)
3. Shen, Y.X., Lin, J., Ji, Z.C.: Study on induction motor backstepping method based on neural network flux estimator. Control Decis. **21**(7), 833–6 (2006)
4. Wang, J.-J., Zhao, G.-Z., Qi, D.-L.: Speed tracking control of permanent magnet synchronous motor with backstepping. Proc. CSEE **24**(8), 95–8 (2004)
5. Wang, L.X., Mendel, J.M.: Fuzzy basis functions, universal approximation, and orthogonal least-squares learning. IEEE Trans. Neural Netw. **3**(5), 807–814 (1992)
6. Yu, H.-S., Zhao, K.-Y., Guo, L., Wang, H.-L.: Maximum torque per ampere control of PMSM based on port-controlled Hamiltonian theory. Proc. CSEE **26**(8), 82–7 (2006)
7. Zhang, C.F., Wang, Y.N., He, J.: Variable structure intelligent control for PM synchronous servo motor drive. Proc. CSEE **22**(7), 13–17 (2002)
8. Zhou, J., Wang, Y.: Adaptive backstepping speed controller design for a permanent magnet synchronous motor. IEE Proc.-Electr. Power Appl. **146**(2), 165–172 (2002)

9. Dou, X., Wang, Y.: Nonlinear golden-section adaptive control of permanent magnet synchronous. J. Syst. Sci. Math. Sci. **35**(7), 860–870 (2015)
10. Du, R., Tao, C., Zhang, W., Zhang, J., Sun, J.: Adaptive fuzzy control method for mechanical resonance suppression of servo systems. Electr. Mach. Control **21**(10), 116–122 (2017)
11. Du, R., Wu, Y., Chen, W., Chen, Q.: Adaptive fuzzy control for the servo system with LuGre friction. Control Decis. **28**(8), 1253–1256 (2013)
12. Fu, P., Chen, Z., Cong, B., Zhao, J.: A position servo system of permanent magnet synchronous motor based on back-stepping adaptive sliding mode control. Trans. China Electrotech. Soc. **28**(9), 288 (2013)
13. Li, C., Chen, M., Han, Y.: Design of position servo system based on maximum phase margin. Trans. China Electrotech. Soc. **30**(20), 10–20 (2015)
14. Li, N., Li, Y., Wang, H., Sun, Y.: Fuzzy tracking control for fractional-order permanent magnet synchronous motor chaotic system. Inf. Control **45**(1), 8–13 (2016)
15. Wang, W., Yu, Y.: Speed tracking control of permanent magnet synchronous motors. J. Syst. Sci. Math. Sci. **35**(9), 1028–1036 (2015)
16. Yu, J., Yu, H., Lin, C.: Fuzzy approximation-based adaptive command filtered backstepping control for induction motors with iron losses. Control Decis. **31**(12), 2189–2194 (2016)
17. Yu, Y., Wang, W.: Adaptive neural networks dynamic surface control for permanent magnet synchronous motor. Comput. Simul. **31**(10), 401 (2014)
18. Zhang, Z., Zhang, T.: The sliding mode of permanent magnet synchronous motor speed controller simulation modeling research. Comput. Simul. **33**(12), 380–384 (2016)
19. Zadeh, L.A.: Fuzzy sets. Inf. Control **8**(3), 338–353 (1965)
20. Tong, S,C., Li, H.H.: Direct adaptive fuzzy output tracking control of nonlinear systems. Fuzzy Sets Syst. **128**, 107–115 (2002)
21. Pillay, P., Krishnan, R.: Modeling of permanent magnet motor drives. IEEE Trans. Ind. Electron. **35**(4), 537–541 (1998)

Chapter 9
Neural Networks-Based Adaptive DSC for PMSM

This chapter considers the problem of neural networks (NNs)-based adaptive dynamic surface control (DSC) for permanent magnet synchronous motor (PMSM) with load torque disturbance and parameter uncertainties. First, neural networks are used to approximate the unknown nonlinear functions of PMSM drive system and a novel adaptive DSC is constructed to avoid the "explosion of complexity" problem existing in the traditional backstepping design. Next, under the proposed adaptive neural DSC, only one adaptive parameter is required, and the designed neural controllers structure is much simpler than some existing results in literature, which can guarantee that the tracking error converges to a small neighborhood of the origin. Finally, The simulation results show the effectiveness and potential of the new design technique.

9.1 Introduction

In the past few decades, permanent magnet synchronous motor (PMSM) has attracted much attention due to its extensive industrial application [1–3]. However, it is still a challenging problem to control the PMSM to obtain ideal dynamic performance because their dynamics are usually multivariable [4–8], highly nonlinear [9] and coupled; and very sensitive to external load disturbances and parameter changes [10, 11]. To achieve better performance of PMSM, many efforts have been devoted to the development of nonlinear control methods for PMSM, and various algorithms have been proposed, see for example [12–18].

In this chapter, a neural networks-based adaptive DSC is proposed to solve the problems of the conventional backstepping method for PMSM drive systems [19, 20]. The RBF networks are used to approximate the unknown nonlinear functions to solve the first problem of "linear in the unknown system parameters" [21], and a DSC

J. Yu et al., *Intelligent Backstepping Control for the Alternating-Current Drive Systems*, Studies in Systems, Decision and Control 349,
https://doi.org/10.1007/978-3-030-67723-7_9

technique is proposed to solve the second problem of "explosion of complexity" by first-order filtering technique [22]. The proposed control scheme not only ensures the boundedness of all signals in the closed-loop system, but also reduces the number of adaptive parameters which alleviates the computational burden [23, 24]. Finally, the effectiveness and robustness of the new design method are verified by simulations [25].

9.2 Mathematical Model of the PMSM Drive System and Preliminaries

From (1.16), the dynamic model of the PMSM can be described as follows:

$$\begin{aligned}
\dot{x}_1 &= x_2, \\
\dot{x}_2 &= \frac{a_1}{J}x_3 + \frac{a_2}{J}x_3x_4 - \frac{B}{J}x_2 - \frac{T_L}{J}, \\
\dot{x}_3 &= b_1x_3 + b_2x_2x_4 + b_3x_2 + b_4u_q, \\
\dot{x}_4 &= c_1x_4 + c_2x_2x_3 + c_3u_d.
\end{aligned} \tag{9.1}$$

In this chapter, the RBF neural networks will be used to approximate the unknown continuous function $\varphi(z) : R^q \to R$ as $\hat{\varphi}(z) = \phi^{*T}P(z)$, where $z \in \Omega_z \subset R^q$ is the input vector with q being the neural networks input dimension, $\phi^* = [\Phi_1^*, ..., \Phi_n^*]^T \in R^n$ is the weight vector, $n > 1$ is the neural networks node number, and $P(z) = [p_1(z), ..., p_n(z)]^T \in R^n$ is the basis function vector with $p_i(z)$ chosen as the commonly used Gaussian function in the following form: $p_i(z) = \exp\left[\frac{-(z-\nu_i)^T(z-\nu_i)}{q_i^2}\right]$, $i = 1, 2, ..., n$ where $\nu_i = [\nu_{i1}, ..., \nu_{iq}]^T$ is the center of the receptive field and q_i is the width of the Gaussian function.

Lemma 9.1 *[21] For a given scalar $\varepsilon > 0$, by choosing sufficiently large l, the RBF neural networks can approximate any continuous function over a compact set $\Omega_z \in R^q$ to an arbitrary accuracy as $\varphi(z) = \phi^T P(z) + \delta(z)\ \forall\ z \in \Omega_z \subset R^q$ where $\delta(z)$ is the approximation error satisfying $|\delta(z)| \le \varepsilon$ and ϕ is an unknown ideal constant weight vector, which is an artificial quantity required for analytical purpose. Typically, ϕ is chosen as the value of ϕ^*that minimizes$|\delta(z)|$ for all $z \in \Omega_z$,i.e.,*

$$\phi := \arg\min_{\phi^* \in R^n} \left\{ \sup_{z \in \Omega_z} \left|\varphi(z) - \phi^{*T}P(z)\right| \right\}.$$

9.3 Adaptive Dynamic Surface Control for PMSM

In this section, An adaptive dynamic surface control for PMSM based on backstepping will be proposed.

Step 1: For the reference signal x_d, we define the tracking error variable as $z_1 = x_1 - x_d$. From the first subsystem of (9.1), the error dynamic system is computed by $\dot{z}_1 = x_2 - \dot{x}_d$.

Choose a Lyapunov function candidate as $V_1 = \frac{1}{2}z_1^2$, then the time derivative of V_1 is given by

$$\dot{V}_1 = z_1\dot{z}_1 = z_1(x_2 - \dot{x}_d). \tag{9.2}$$

Construct the virtual control law α_1 as

$$\alpha_1 = -k_1 z_1 + \dot{x}_d, \tag{9.3}$$

with $k_1 > 0$ being a design control gain. Next, introduce a new state variable α_{1d}. Let α_1 pass through a first-order filter with time constant ϵ_1 to obtain α_{1d} as

$$\epsilon_1\dot{\alpha}_{1d} + \alpha_{1d} = \alpha_1, \alpha_{1d}(0) = \alpha_1(0). \tag{9.4}$$

Define $z_2 = x_2 - \alpha_{1d}$. By using (9.3) and (9.4), (9.2) can be rewritten in the following form.

$$\dot{V}_1 = z_1\dot{z}_1 = z_1(z_2 + \alpha_{1d} - \dot{x}_d) = -k_1 z_1^2 + z_1 z_2 + z_1(\alpha_{1d} - \alpha_1). \tag{9.5}$$

Step 2: Differentiating z_2 obtains

$$\dot{z}_2 = \frac{a_1}{J}x_3 + \frac{a_2}{J}x_3 x_4 - \frac{B}{J}x_2 - \frac{T_L}{J} - \dot{\alpha}_{1d}. \tag{9.6}$$

Now, choose the Lyapunov function candidate as $V_2 = V_1 + \frac{J}{2}z_2^2$. Obviously, the time derivative of V_2 can be expressed as

$$\dot{V}_2 = -k_1 z_1^2 + z_1(\alpha_{1d} - \alpha_1) + z_2(a_1 x_3 + f_2), \tag{9.7}$$

where $f_2(Z_2) = z_1 + a_2 x_3 x_4 - Bx_2 - T_L - J\dot{\alpha}_{1d}$ and $Z_2 = [x_1, x_2, x_3, x_4, x_d, \dot{x}_d]^T$.

Remark 9.2 It should be pointed that the system parameters B, T_L and J may be unknown in the PMSM drive system, then they cannot be used to construct the control signal unless we specify their corresponding adaptation laws. To avoid this trouble, we will employ the neural networks to approximate the nonlinear function f_2, and the indeterministic parameters will be taken into account. ♦

According to the RBF neural networks approximation property, for a given $\varepsilon_2 > 0$, there exists a RBF neural networks $\phi_2^T P_2(Z_2)$ such that $f_2 = \phi_2^T P_2(Z_2) + \delta_2(Z_2)$ where $\delta_2(Z_2)$ is the approximation error satisfying $|\delta_2| \leq \varepsilon_2$. Consequently, we can show the following inequality:

$$z_2 f_2 = z_2\left(\phi_2^T P_2 + \delta_2\right) \leq \frac{1}{2l_2^2}z_2^2 \|\phi_2\|^2 P_2^T P_2 + \frac{1}{2}(l_2^2 + z_2^2 + \varepsilon_2^2). \tag{9.8}$$

Then we construct the virtual control α_2 as

$$\alpha_2 = \frac{1}{a_1}(-k_2 z_2 - \frac{1}{2} z_2 - \frac{1}{2l_2^2} z_2 \hat{\theta} P_2^T P_2), \tag{9.9}$$

where $\hat{\theta}$ is the estimation of θ, the unknown constant θ which will be specified later. Define a new state variable α_{2d}. Let α_2 passes through a first-order filter with the time constant ϵ_2 to obtain α_{2d} as

$$\epsilon_2 \dot{\alpha}_{2d} + \alpha_{2d} = \alpha_2, \alpha_{2d}(0) = \alpha_2(0). \tag{9.10}$$

and define $z_3 = x_3 - \alpha_{2d}$, then we can get

$$\begin{aligned} \dot{V}_2 = &- \sum_{i=1}^{2} k_i z_i^2 + z_1(\alpha_{1d} - \alpha_1) + a_1 z_2(\alpha_{2d} - \alpha_2) + \frac{1}{2} l_2^2 \\ &+ \frac{1}{2l_2^2} z_2^2 (\|\phi_2\|^2 - \hat{\theta}) P_2^T P_2 + a_1 z_2 z_3 + \frac{1}{2} \varepsilon_2^2. \end{aligned} \tag{9.11}$$

Step 3: Differentiating z_3 obtains

$$\dot{z}_3 = \dot{x}_3 - \dot{\alpha}_{2d} = b_1 x_3 + b_2 x_2 x_4 + b_3 x_2 + b_4 u_q - \dot{\alpha}_{2d}.$$

Choose the following Lyapunov function candidate as $V_3 = V_2 + \frac{1}{2} z_3^2$. Then, differentiating V_3 yields

$$\begin{aligned} \dot{V}_3 = &- \sum_{i=1}^{2} k_i z_i^2 + z_1(\alpha_{1d} - \alpha_1) + a_1 z_2(\alpha_{2d} - \alpha_2) + \frac{1}{2} l_2^2 \\ &+ \frac{z_2^2}{2l_2^2} (\|\phi_2\|^2 - \hat{\theta}) P_2^T P_2 + \frac{1}{2} \varepsilon_2^2 + z_3(f_3 + b_4 u_q), \end{aligned} \tag{9.12}$$

where $f_3(Z_3) = b_1 x_3 + b_2 x_2 x_4 + b_3 x_2 + a_1 z_2 - \dot{\alpha}_{2d}$, $Z_3 = Z_2$.

Remark 9.3 It should be noted that f_3 contains the derivative of α_{2d} and the nonlinear term $b_2 x_2 x_4$, this will make the backstepping design become very difficult, and the designed u_q will have a complex structure. To solve this problem, we will use neural networks to approximate the nonlinear function f_3. ♦

Similarly, for given $\varepsilon_3 > 0$, there exists $\phi_3^T P_3(Z_3)$ such that

$$z_3 f_3 \leq \frac{1}{2l_3^2} z_3^2 \|\phi_3\|^2 P_3^T P_3 + \frac{1}{2} l_3^2 + \frac{1}{2} z_3^2 + \frac{1}{2} \varepsilon_3^2. \tag{9.13}$$

Thus, by substituting (9.13) into (9.12) it follows that

$$\dot{V}_3 \le -\sum_{i=1}^{2} k_i z_i^2 + z_1(\alpha_{1d} - \alpha_1) + a_1 z_2(\alpha_{2d} - \alpha_2) + \frac{1}{2l_2^2} z_2^2(\|\phi_2\|^2 - \hat{\theta}) P_2^T P_2$$
$$+ \frac{1}{2l_3^2} z_3^2 \|\phi_3\|^2 P_3^T P_3 \sum_{i=2}^{3} \frac{1}{2} l_i^2 + \sum_{i=2}^{3} \frac{1}{2}\varepsilon_i^2 + \frac{1}{2} z_3^2 + z_3 b_4 u_q.$$

The control input u_q is designed as

$$u_q = \frac{1}{b_4}(-k_3 z_3 - \frac{1}{2} z_3 - \frac{1}{2l_3^2} z_3 \hat{\theta} P_3^T P_3). \tag{9.14}$$

Furthermore, using equality (9.14), we can get

$$\dot{V}_3 \le -\sum_{i=1}^{3} k_i z_i^2 + z_1(\alpha_{1d} - \alpha_1) + a_1 z_2(\alpha_{2d} - \alpha_2)$$
$$+ \sum_{i=2}^{3} \frac{1}{2l_i^2} z_i^2(\|\phi_i\|^2 - \hat{\theta}) P_i^T P_i + \sum_{i=2}^{3} \frac{1}{2} l_i^2 + \sum_{i=2}^{3} \frac{1}{2}\varepsilon_i^2.$$

Step 4: At this step, we will construct the control law u_d. Define $z_4 = x_4$ and choose $V_4 = V_3 + \frac{1}{2} z_4^2$. Then differentiating V_4 yields

$$\dot{V}_4 \le -\sum_{i=1}^{3} k_i z_i^2 + z_1(\alpha_{1d} - \alpha_1) + a_1 z_2(\alpha_{2d} - \alpha_2)$$
$$+ \sum_{i=2}^{3} \frac{1}{2l_i^2} z_i^2(\|\phi_i\|^2 - \hat{\theta}) P_i^T P_i + \sum_{i=2}^{3} \frac{1}{2} l_i^2 + \sum_{i=2}^{3} \frac{1}{2}\varepsilon_i^2$$
$$+ z_4\left(f_4(Z_4) + c_3 u_d\right), \tag{9.15}$$

where $f_4(Z_4) = c_1 z_4 + c_2 x_2 x_3$ and $Z_4 = Z_2$. Similarly, for a given $\varepsilon_4 > 0$, there exists $\phi_4^T P_4(Z_4)$ satisfying

$$z_4 f_4(Z_4) \le \frac{1}{2l_4^2} z_4^2 \|\phi_4\|^2 P_4^T P_4 + \frac{1}{2} l_4^2 + \frac{1}{2} z_4^2 + \frac{1}{2}\varepsilon_4^2. \tag{9.16}$$

Substituting (9.16) into (9.15) gives

$$\dot{V}_4 \le -\sum_{i=1}^{3} k_i z_i^2 + z_1(\alpha_{1d} - \alpha_1) + a_1 z_2(\alpha_{2d} - \alpha_2)$$
$$+ \sum_{i=2}^{3} \frac{1}{2l_i^2} z_i^2(\|\phi_i\|^2 - \hat{\theta}) P_i^T P_i + \sum_{i=2}^{4} \frac{1}{2} l_i^2 + \sum_{i=2}^{4} \frac{1}{2}\varepsilon_i^2$$

$$+\frac{1}{2l_4^2}z_4^2\left\|\phi_4\right\|^2 P_4^T P_4 + \frac{1}{2}z_4^2 + c_3 z_4 u_d. \tag{9.17}$$

u_d is designed as

$$u_d = \frac{1}{c_3}(-k_4 z_4 - \frac{1}{2}z_4 - \frac{1}{2l_4^2}z_4\hat{\theta}P_4^T P_4). \tag{9.18}$$

Define $\theta = \max\{\left\|\phi_2\right\|^2, \left\|\phi_3\right\|^2, \left\|\phi_4\right\|^2\}$. Furthermore, combining (9.17) with (9.18) results in

$$\begin{aligned}\dot{V}_4 \le& -\sum_{i=1}^{4}k_i z_i^2 + z_1(\alpha_{1d}-\alpha_1) + a_1 z_2(\alpha_{2d}-\alpha_2)\\&+\sum_{i=2}^{4}\frac{1}{2}(l_i^2+\varepsilon_i^2) + \sum_{i=2}^{4}\frac{1}{2l_i^2}z_i^2 P_i^T P_i\left(\theta-\hat{\theta}\right).\end{aligned} \tag{9.19}$$

Introduce y_1, y_2 and $\tilde{\theta}$ as

$$y_1 = \alpha_{1d}-\alpha_1,\ y_2 = \alpha_{2d}-\alpha_2,\ \tilde{\theta} = \hat{\theta}-\theta. \tag{9.20}$$

Then we can get

$$\begin{aligned}\dot{y}_1 &= \dot{\alpha}_{1d}-\dot{\alpha}_1 = -\frac{\alpha_{1d}-\alpha_1}{\epsilon_1} - \dot{\alpha}_1\\&= -\frac{y_1}{\epsilon_1} + k_1\dot{z}_1 - \ddot{x}_d = -\frac{y_1}{\epsilon_1} + D_1,\\\dot{y}_2 &= -\frac{y_2}{\epsilon_2} + D_2,\end{aligned} \tag{9.21}$$

with $D_2 = \frac{1}{a_1}(k_2 + \frac{1}{2} + \frac{1}{2l_2^2}\hat{\theta}P_2^T P_2)\dot{z}_2 + \frac{1}{2a_1 l_2^2}\dot{\hat{\theta}}P_2^T P_2 z_2 + \frac{1}{a_1 l_2^2}\hat{\theta}(\dot{Z}_2\frac{\partial P_2(Z_2)}{\partial Z_2}P_2(Z_2)z_2$.
Choose the following Lyapunov function candidate:

$$V = V_4 + \frac{1}{2}y_1^2 + \frac{1}{2}y_2^2 + \frac{1}{2r_1}\tilde{\theta}^2, \tag{9.22}$$

where r_1 is a positive constant. By differentiating V and taking (9.19)–(9.22) into account, one has

$$\begin{aligned}\dot{V} \le& -\sum_{i=1}^{4}k_i z_i^2 + \sum_{i=2}^{4}\frac{1}{2}(l_i^2+\varepsilon_i^2) + \sum_{i=1}^{2}y_i\dot{y}_i + z_1 y_1\\&+a_1 z_2 y_2 + \frac{1}{r_1}\tilde{\theta}\left[-\sum_{i=2}^{4}\frac{r_1}{2l_i^2}z_i^2 P_i^T P_i + \dot{\hat{\theta}}\right].\end{aligned} \tag{9.23}$$

According to (9.23), the corresponding adaptive law is chosen as follows:

$$\dot{\hat{\theta}} = \sum_{i=2}^{4} \frac{r_1}{2l_i^2} z_i^2 P_i^T P_i - m_1 \hat{\theta}, \tag{9.24}$$

where m_1 and l_i for $i = 2, 3, 4$ are positive constants.

Remark 9.4 By combining the RBF networks approximation and DSC technique, the controller designed has a simpler structure and the problems of "linear in the unknown system parameters" and "explosion of complexity" are overcome. In addition, the number of adaptive parameters is reduced to only one, while four adaptive parameters are required in [22]. This will alleviate the computational burden and render the designed scheme more effective and suitable in practical applications. ♦

Remark 9.5 It can be clearly seen that the proposed controllers (9.14) and (9.18) have a simpler structure. This means that the proposed neural networks-based adaptive dynamic surface controllers are easy to be implemented in real world applications. ♦

Theorem 9.6 *Consider system (9.1) and the given reference signal x_d. Then under the action of the neural networks-based adaptive dynamic surface controllers (9.14), (9.18) and the adaptive law (9.24), the tracking error of the closed-loop controlled system will converge to a sufficiently small neighborhood of the origin and all the closed-loop signals will be bounded.*

Proof To address the stability analysis of the resulting closed-loop system, by substituting (9.24) into (9.23), one has

$$\begin{aligned} \dot{V} \le & -\sum_{i=1}^{4} k_i z_i^2 + \sum_{i=2}^{4} \frac{1}{2}(l_i^2 + \varepsilon_i^2) + \sum_{i=1}^{2} y_i \dot{y}_i \\ & + z_1 y_1 + a_1 z_2 y_2 - \frac{m_1}{r_1} \tilde{\theta} \hat{\theta}. \end{aligned} \tag{9.25}$$

According to [23, 24], $|D_i|$ has a maximum D_{iM} on compact set $|\Omega_i|$, $i = 1, 2$, i.e., $|D_i| \le D_{iM}$. Therefore, we can get

$$\begin{aligned} y_1 \dot{y}_1 &\le -\frac{y_1^2}{\epsilon_1} + |D_{1M}||y_1| \le -\frac{y_1^2}{\epsilon_1} + \frac{1}{2\tau} D_{1M}^2 y_1^2 + \frac{\tau}{2}, \\ y_2 \dot{y}_2 &\le -\frac{y_2^2}{\epsilon_2} + \frac{1}{2\tau} D_{2M}^2 y_2^2 + \frac{\tau}{2}, \end{aligned}$$

with $\tau > 0$. Using the following inequalities $z_1 y_1 \le \frac{1}{4} y_1^2 + z_1^2$, $a_1 z_2 y_2 \le \frac{a_1^2}{4} y_2^2 + z_2^2$ and $-\tilde{\theta}\hat{\theta} \le -\tilde{\theta}(\tilde{\theta} + \theta) \le -\frac{1}{2}\tilde{\theta}^2 + \frac{1}{2}\theta^2$, (9.25) can be rewritten in the following form:

$$\dot{V} \le -(k_1-1)z_1^2-(k_2-1)z_2^2-\sum_{i=3}^{4}k_iz_i^2-\frac{m_1}{2r_1}\tilde{\theta}^2$$
$$-(\frac{1}{\epsilon_1}-(\frac{1}{4}+\frac{1}{2\tau}D_{1M}^2))y_1^2+\sum_{i=2}^{4}\frac{1}{2}(l_i^2+\varepsilon_i^2)+\frac{m_1}{2r_1}\theta^2$$
$$-(\frac{1}{\epsilon_2}-(\frac{a_1^2}{4}+\frac{1}{2\tau}D_{2M}^2))y_2^2+\tau. \tag{9.26}$$

Choose the design parameters k_1, k_2 and τ such that $k_1-1>0$, $k_2-1>0$, $\frac{1}{\epsilon_1}-(\frac{1}{4}+\frac{1}{2\tau}D_{1M}^2)>0$ and $\frac{1}{\epsilon_2}-(\frac{a_1^2}{4}+\frac{1}{2\tau}D_{2M}^2)>0$. Also we can obtain $\dot{V}\le -a_0V+b_0$ where $a_0=\min\Big\{2(k_1-1),\ \frac{2(k_2-1)}{J},\ 2k_3,\ 2(\frac{1}{\epsilon_1}-(\frac{1}{4}+\frac{1}{2\tau}D_{1M}^2)),\ 2k_4,\ m_1,\ 2(\frac{1}{\epsilon_2}-(\frac{a_1^2}{4}+\frac{1}{2\tau}D_{2M}^2))\Big\}$ and $b_0=\sum_{i=2}^{4}\frac{1}{2}(l_i^2+\varepsilon_i^2)+\frac{m_1}{2r_1}\theta^2+\tau$. Furthermore, the above inequality implies that

$$V(t)\le (V(t_0)-\frac{b_0}{a_0})e^{-a_0(t-t_0)}+\frac{b_0}{a_0}\le V(t_0)+\frac{b_0}{a_0},\ \forall t\ge t_0. \tag{9.27}$$

As a result, all $z_i(i=1,2,3,4)$, $y_i(i=1,2)$ and $\tilde{\theta}$ belong to the compact set $\Omega=\Big\{(z_i,y_1,y_2,\tilde{\theta})|V\le V(t_0)+\frac{b_0}{a_0},\ \forall t\ge t_0\Big\}$. Namely, all the signals in the closed-loop system are bounded. Especially, from (9.27) we have

$$\lim_{t\to\infty} z_1^2 \le \frac{2b_0}{a_0}. \tag{9.28}$$

Remark 9.7 It can be seen from the definitions of a_0 and b_0 that to get a small tracking error we can set r_1 large, but l_i and ε_i small enough after giving the parameters k_i, ϵ_i, τ and m_1. ♦

■

9.4 A Comparison with the Classical Backstepping Design

9.4.1 Classical Backstepping Design

This part is devoted to provide the designed controllers by classical backstepping approach in [22].

$$
\begin{aligned}
u_q = & \frac{1}{b_4}(-k_3 z_3 - b_1 x_3 - b_3 x_2 - a_1 z_2 + \frac{\partial \alpha_2}{\partial x_1} x_2 \\
& + \sum_{i=0}^{2} \frac{\partial \alpha_2}{\partial x_d^{(i)}} x_d^{(i+1)} + \frac{\partial \alpha_2}{\partial \hat{B}} \dot{\hat{B}} + \frac{\partial \alpha_2}{\partial \hat{T}_L} \dot{\hat{T}}_L + \frac{\partial \alpha_2}{\partial \hat{J}} \dot{\hat{J}}) \\
& + \frac{\partial \alpha_2}{\partial x_2} \left(\frac{a_1}{J} x_3 + \frac{a_2}{J} x_3 x_4 - \frac{B}{J} x_2 - \frac{T_L}{J} \right),
\end{aligned} \tag{9.29}
$$

$$
u_d = -\frac{1}{c_3}(k_4 z_4 + a_2 z_2 x_3 + b_2 x_2 z_3 + c_1 z_4 + c_2 x_2 x_3). \tag{9.30}
$$

Remark 9.8 By comparing the neural networks-based adaptive dynamic surface controllers (9.14) and (9.18) with the classical backstepping controllers (9.29) and (9.30) given in [22], it can be seen that the classical controllers (9.29) and (9.30) are much more complicated than the dynamics surface controllers (9.14) and (9.18). The number of terms in classical backstepping controllers is much larger. This drawback was called the "explosion of terms" in [25].

9.4.2 Simulation Results

In this section, an example is used to make a comparison between the proposed neural networks-based adaptive dynamic surface controllers (9.14) and (9.18) and the classical backstepping controllers (9.29) and (9.30) given in [22] for the PMSM drive system with the following parameters:

$$
\begin{aligned}
& J = 0.00379\,\text{Kgm}^2, R_s = 0.68\Omega, L_d = 0.00315\text{H}, n_p = 3, \\
& L_q = 0.00285\text{H}, B = 0.001158\,\text{Nm/(rad/s)}, \Phi = 0.1245\text{H}.
\end{aligned}
$$

The simulation is carried out under the zero initial condition. The reference signals are chosen as $x_d = 0.5\sin(t) + \sin(0.5t)$ and $T_L = \begin{cases} 1.5, & 0 \leq t \leq 20, \\ 3, & t \geq 20. \end{cases}$ The RBF neural networks are chosen in the following way. Neural networks $\phi_2^T P_2(Z_2)$, $\phi_3^T P_3(Z_3)$ and $\phi_4^T P_4(Z_4)$ contain eleven nodes with centers spaced evenly in the interval $[-10, 10]$ and widths being equal to 2, respectively. The proposed adaptive neural controllers in this paper are used to control this PMSM motor. The control parameters are chosen as follows:

$$
\begin{aligned}
& k_1 = 60, \quad k_2 = 20, \quad k_3 = 35, \quad k_4 = 25, \quad r_1 = 0.01, \\
& m_1 = 0.05, \quad l_2 = l_3 = l_4 = 0.5.
\end{aligned}
$$

The classical backstepping controllers (9.29) and (9.30) are also utilized to control the systems and the controller parameters $k_i (i = 1, 2, 3, 4)$ are chosen as the same as the those in the above adaptive neural controllers.

The simulation results for the above two control methods are shown in Figs. 9.1, 9.2, 9.3, 9.4, 9.5, 9.6, 9.7, 9.8, 9.9 and 9.10. Note that Figs. 9.1, 9.2, 9.3 and 9.4 display

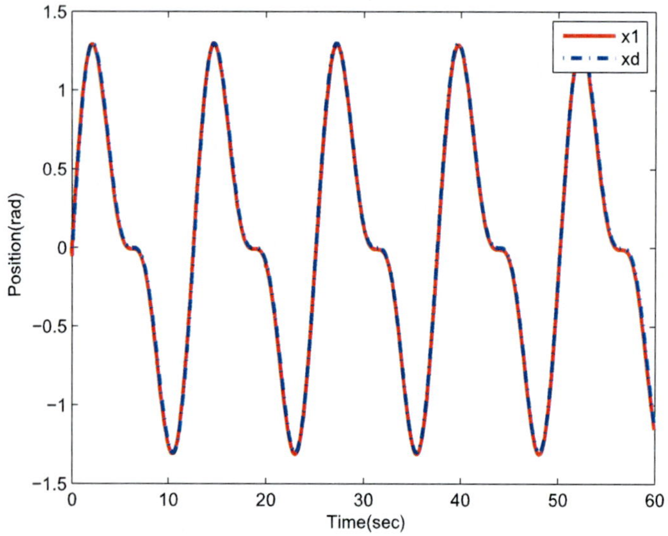

Fig. 9.1 Trajectories of the x_1 and x_d for dynamic surface control

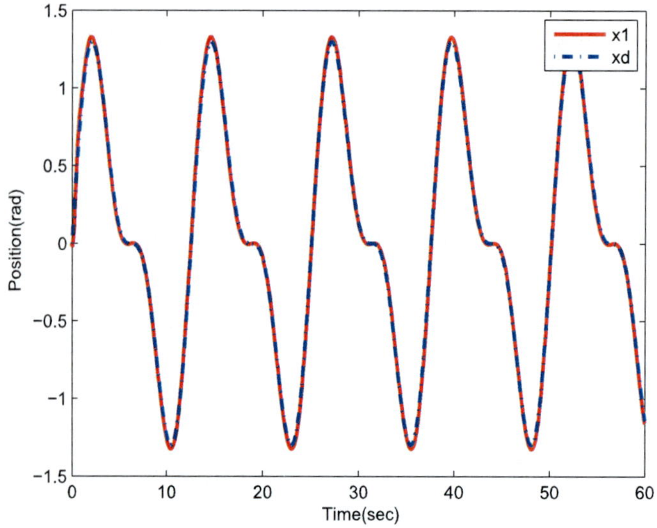

Fig. 9.2 Trajectories of the x_1 and x_d for classical backstepping

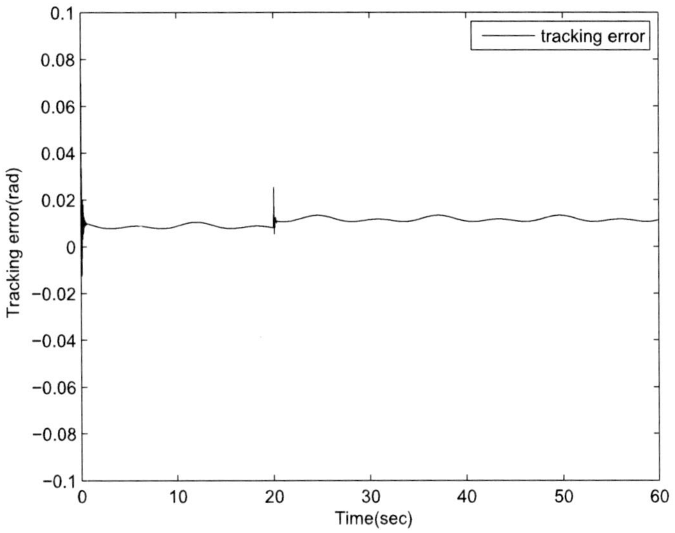

Fig. 9.3 The tracking error of x_1 and x_d for dynamic surface control

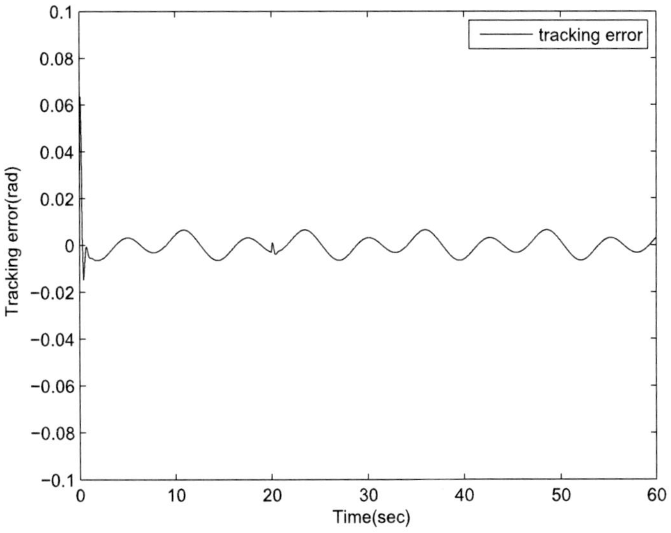

Fig. 9.4 The tracking error of x_1 and x_d for classical backstepping

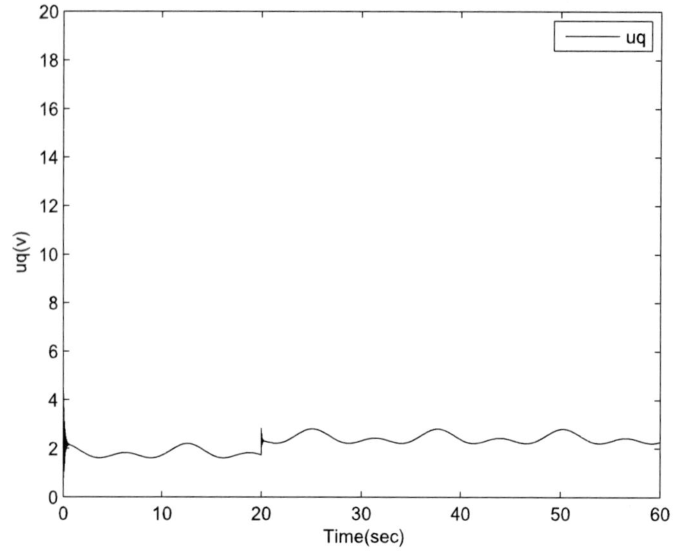

Fig. 9.5 Curve of the u_q for dynamic surface control

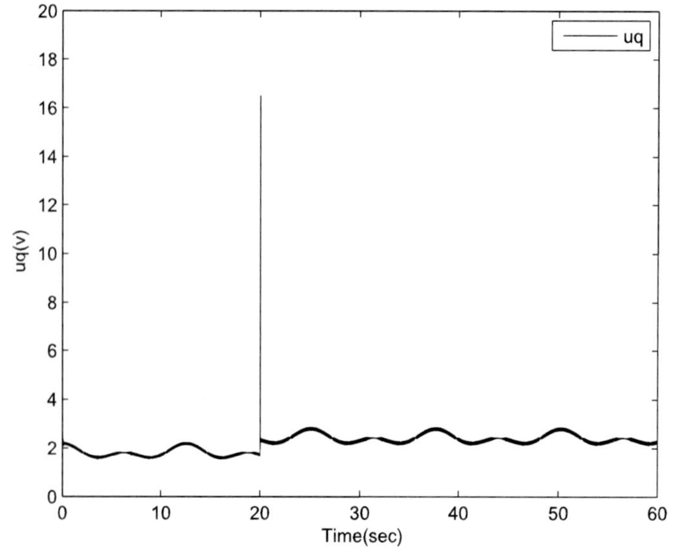

Fig. 9.6 Curve of the u_q for classical backstepping

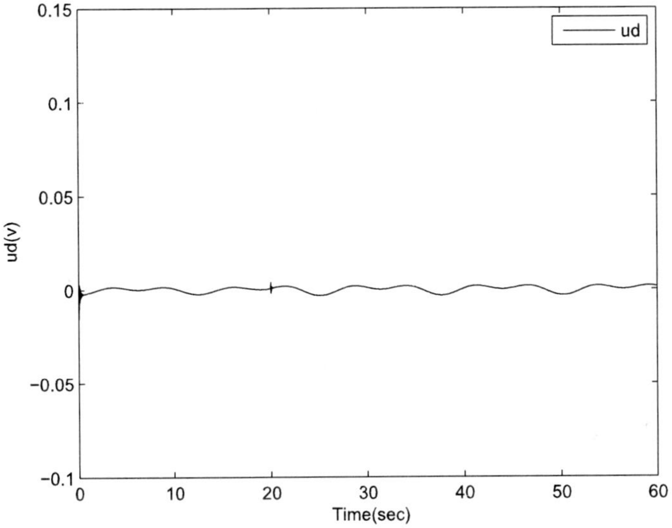

Fig. 9.7 Curve of the u_d for dynamic surface control

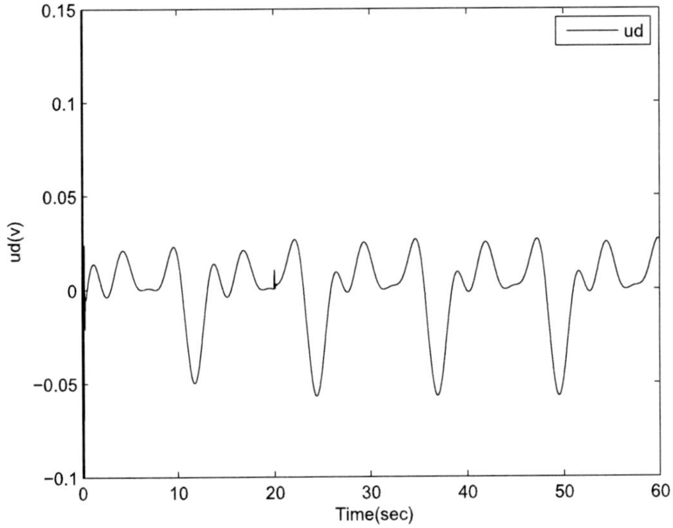

Fig. 9.8 Curve of the u_d for classical backstepping

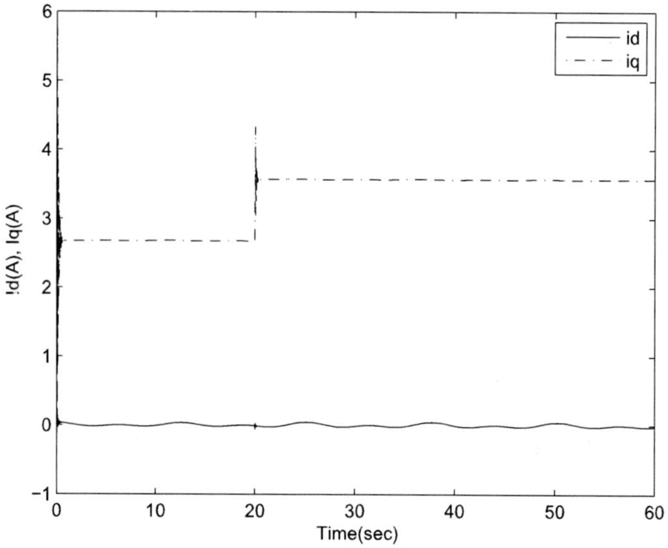

Fig. 9.9 Curves of the i_d, i_q for dynamic surface control

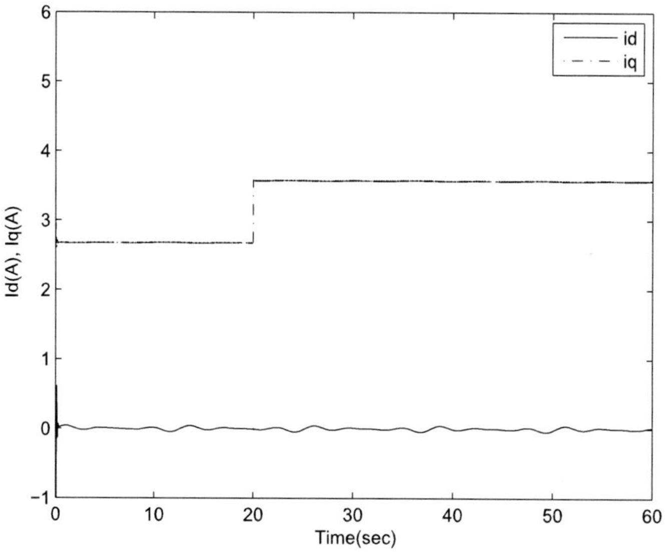

Fig. 9.10 Curves of the i_d, i_q for classical backstepping

the system outputs, the reference signals and the tracking error for both control approaches. From Figs. 9.1, 9.2, 9.3 and 9.4, it can be clearly seen that under the actions of controllers (9.14) and (9.18) and the traditional backstepping controllers (9.29) and (9.30) in [22], the system outputs follow the desired reference signals well. The control input signals are shown in Figs. 9.5, 9.6, 9.7 and 9.8; while Figs. 9.9 and 9.10 display the trajectories of i_d and i_q. From the simulations, it is clearly shown that the proposed adaptive neural dynamics surface controllers in this paper can trace the reference signal quite well, even though the controllers have much simpler structure than the classical ones, which is more practical to be implemented.

9.5 Conclusion

In this chapter, the neural networks-based adaptive dynamic surface control is designed for PMSM. The proposed control method can overcome not only the problem of "linear in the unknown system parameters", but also the "explosion of complexity" inherent in the backstepping design. It is demonstrated that under such controllers, the boundedness of all signals in the closed-loop system can be guaranteed, and the tracking error can converge to a small neighborhood of the origin. The effectiveness and robustness of the developed new control scheme against the parameter uncertainties and load disturbances are illustrated by an example and simulation.

References

1. Rezaei, M., Asadizadeh, M.: Predicting unconfined compressive strength of intact rock using new hybrid intelligent models. J. Mining Environ. **11**(1), 231–246 (2019)
2. Gao, W., Su, C.: Analysis on block chain financial transaction under artificial neural network of deep learning. J. Comput. Appl. Math. **380** (2020)
3. Sun, K., Mou, S., Qiu, J., Wang, T., Gao, H.: Adaptive fuzzy control for nontriangular structural stochastic switched nonlinear systems with full state constraints. IEEE Trans. Fuzzy Syst. **27**(8), 1587–1601 (2019)
4. Xia, J., Zhang, J., Sun, W., Zhang, B., Wang, Z.: Finite-time adaptive fuzzy control for nonlinear systems with full state constraints. IEEE Trans. Syst. Man Cybern. Syst. **49**(7), 1541–1548 (2019)
5. Zhang, Z., Liang, H., Wu, C., Ahn, C.: Adaptive event-triggered output feedback fuzzy control for nonlinear networked systems with packet dropouts and actuator failure. IEEE Trans. Fuzzy Syst. **27**(9), 1793–1806 (2019)
6. Zhao, X., Wang, X., Zhang, S., Zong, G.: Adaptive neural backstepping control design for a class of nonsmooth nonlinear systems. IEEE Trans. Syst. Man Cybern. **49**(9), 1820–1831 (2019)
7. Leonhard, W.: Control of Electrical Drives (1985)
8. Wai, R.: Total sliding-mode controller for PM synchronous servo motor drive using recurrent fuzzy neural network. IEEE Trans. Ind. Electron. **48**(5), 926–944 (2001)
9. Shang, W., Zhao, S., Shen, Y., Qi, Z.: A sliding mode flux-linkage controller with integral compensation for switched reluctance motor. IEEE Trans. Magn. **45**(9), 3322–3328 (2009)

10. Verrelli, C.: Adaptive learning control design for robotic manipulators driven by permanent magnet synchronous motors. Int. J. Control **84**(6), 1024–1030 (2011)
11. Verrelli, C.: Synchronization of permanent magnet electric motors: new nonlinear advanced results. Nonlinear Anal. Real World Appl. **13**(1), 395–409 (2012)
12. Guo, Y., Xi, Z., Cheng, D.: Speed regulation of permanent magnet synchronous motor via feedback dissipative Hamiltonian realisation. IET Control Theory Appl. **1**(1), 281–290 (2007)
13. Chen, Z., Tomita, M., Doki, S., Okuma, S.: An extended electromotive force model for sensorless control of interior permanent-magnet synchronous motors. IEEE Trans. Ind. Electron. **50**(2), 288–295 (2003)
14. Hongzhe, J., Jangmyung, L.: An RMRAC current regulator for permanent-magnet synchronous motor based on statistical model interpretation. IEEE Trans. Ind. Electron. **56**(1), 169–177 (2009)
15. Baik, I., Kim, K., Youn, M.: Robust nonlinear speed control of PM synchronous motor using boundary layer integral sliding mode control technique. IEEE Trans. Control Syst. Technol. **8**(1), 47–54 (2000)
16. Zhou, J., Wang, Y.: Real-time nonlinear adaptive backstepping speed control for a PM synchronous motor. Control Eng. Pract. **13**(10), 1259–1269 (2005)
17. Chaoui, H., Sicard, P.: Adaptive fuzzy logic control of permanent magnet synchronous machines with nonlinear friction. IEEE Trans. Ind. Electron. **59**(2), 1123–1133 (2011)
18. Barkat, S., Tlemçani, A., Nouri, H.: Noninteracting adaptive control of PMSM using interval Type-2 fuzzy logic systems. IEEE Trans. Fuzzy Syst. **19**(5), 925–936 (2011)
19. Li, T., Wang, D., Feng, G., Tong, S.: A DSC approach to robust adaptive NN tracking control for strict-feedback nonlinear systems. IEEE Trans. Syst. Man Cybern. Part B Cybern. **40**(3), 915–927 (2010)
20. Li, T., Tong, S., Feng, G.: A novel robust adaptive-fuzzy-tracking control for a class of nonlinear Multi-Input/Multi-Output systems. IEEE Trans. Fuzzy Syst. **18**(1), 150–160 (2010)
21. Sanner, R., Slotine, J.: Gaussian networks for direct adaptive control. IEEE Trans. Neural Netw. **3**(6), 837–863 (1992)
22. Yu, J., Ma, Y., Chen, B., Yu, H.: Adaptive fuzzy backstepping position tracking control for a permanent magnet synchronous motor. Int. J. Innov. Comput. Inf. Control **7**(4), 1589–1601 (2011)
23. Tong, S., Li, Y., Feng, G., Li, T.: Observer-based adaptive fuzzy backstepping dynamic surface control for a class of MIMO nonlinear systems. IEEE Trans. Syst. Man Cybern. Part B Cybern. **41**(4), 1124–1135 (2011)
24. Wang, D., Huang, J.: Neural network-based adaptive dynamic surface control for a class of uncertain nonlinear systems in strict-feedback form. IEEE Trans. Neural Netw. **16**(1), 195–202 (2005)
25. Tong, S., He, X., Zhang, H.: A combined backstepping and small-gain approach to robust adaptive fuzzy output feedback control. IEEE Trans. Fuzzy Syst. **17**(5), 1059–1069 (2009)

Chapter 10
Discrete-Time Adaptive Position Tracking Control for IPMSM

This paper proposed a discrete-time adaptive position tracking control method for interior permanent magnet synchronous motor (IPMSM) based on fuzzy-approximation. The accurate approximate discrete-time IPMSM position tracking system model is derived by direct discretization using the Euler method. Fuzzy logic systems are used to approximate the nonlinearities of the discrete-time IPMSM drive system. Then a discrete-time fuzzy position tracking controller is designed via a backstepping approach. Compared with existing results, the advantage of the proposed scheme is that the number of adaptive parameters is reduced to only two and the problem of coupling nonlinearity can be overcome. The proposed discrete-time fuzzy controller can guarantee the tracking error converges to a small neighborhood of the origin and all the signals are bounded. Simulation results are provided to demonstrate the effectiveness and robustness of the proposed method against the system parameter variations and load disturbances.

10.1 Introduction

In recent years, the interior permanent magnet synchronous motor (IPMSM) has received increased attention for high-performance electric drive applications in virtue of its considerable advantages such as wide speed operation range, high power density, large torque to inertia ratio and free from maintenance [1–4]. In order to achieve better performance of the IPMSM, many researchers devoted to developing nonlinear control methods for the IPMSM and various schemes have been investigated including nonlinear fuzzy logic control [5–7], adaptive backstepping control [11–14]. However, most of those methods above were limited to nonlinear continuous-time systems, while nonlinear discrete-time control design techniques for the PMSM drive system have not been discussed to the same degree. In terms of stability and achiev-

J. Yu et al., *Intelligent Backstepping Control for the Alternating-Current Drive Systems*, Studies in Systems, Decision and Control 349,
https://doi.org/10.1007/978-3-030-67723-7_10

able performances, the discrete-time approach is regarded as typically superior to the continuous-time emulation approach [15], which has motivated an interesting research activity in the design of controllers based on the discrete-time model of the system. Therefore, there already exist good publications about control methods for discrete-time motor systems [16–19].

In this chapter, a discrete-time adaptive position tracking control for IPMSM is proposed based on fuzzy-approximation. Compared with the relevant results for the IPMSM, the main contributions of this paper can be summarized as follows: (1) The Euler direct discretization method is used to obtain the accurate approximate discrete-time IPMSM system model; (2) A discrete-time adaptive fuzzy position tracking controller which overcomes the coupling nonlinearity because of $L_d \neq L_q$ is proposed; (3) The noncausal problem for nonlinear discrete-time IPMSM drive system combined backstepping is overcome, without the need to transform the system model into a predictor form [20]; (4) The number of adaptive parameters is reduced to only two, which reduces the burden of online calculation and is more suitable for practical engineering applications.

10.2 Mathematical Model of the IPMSM Drive System and Preliminaries

In this section, some preparatory knowledge of an IPMSM will be first introduced. To obtain the mathematical model of an IPMSM, the following assumptions are necessary.

Assumption 10.1 *[21] Saturation and iron losses are neglected although it can be taken into account by parameter changes.*

Assumption 10.2 *[21] The back electromotive force is sinusoidal.*

The model of IPMSM in the $d-q$ frame is described as:

$$\begin{aligned}
\dot{\theta}(t) &= \omega(t),\\
\dot{\omega}(t) &= \frac{3n_p\Phi}{2J}i_{qs}(t) - \frac{B}{J}\omega(t) + \frac{3n_p(L_d-L_q)}{2J}i_{ds}(t)\,i_{qs}(t) - \frac{1}{J}T_L,\\
\dot{i}_{qs}(t) &= -\frac{R_s}{L_q}i_{qs}(t) - \frac{n_p\Phi}{L_q}\omega(t) - \frac{n_pL_d}{L_q}\omega(t)\,i_{ds}(t) + \frac{1}{L_q}u_{qs}(t),\\
\dot{i}_{ds}(t) &= -\frac{R_s}{L_d}i_{ds}(t) + \frac{n_p\Phi}{L_d}\omega(t)\,i_{qs}(t) + \frac{1}{L_d}u_{ds}(t),
\end{aligned}$$

where T_L, θ and ω denote the load torque, rotor position and rotor angular velocity. i_{ds} and i_{qs} stand for the $d-q$ axis currents. u_{ds} and u_{qs} are the $d-q$ axis voltages. n_p denotes the pole pairs, the stator resistance R_s, L_d and L_q are the $d-q$ axis stator inductance, the rotor inertia J, the viscous friction coefficient B and the magnetic flux Φ.

By the Euler method, the discrete-time dynamic model of IPMSM drivers can be obtained as follows:

$$\begin{aligned}
\theta(k+1) &= \theta(k) + \Delta_t \omega(k), \\
\omega(k+1) &= \frac{3n_p\Phi}{2J}\Delta_t i_{qs}(k) + (1 - \frac{B}{J}\Delta_t)\omega(k) \\
&\quad + \frac{3n_p(L_d - L_q)}{2J}\Delta_t i_{ds}(k)\, i_{qs}(k) - \frac{1}{J}\Delta_t T_L, \\
i_{qs}(k+1) &= (1 - \frac{R_s}{L_q}\Delta_t) i_{qs}(k) - \frac{n_p\Phi}{L_q}\Delta_t \omega(k) \\
&\quad - \frac{n_p L_d}{L_q}\Delta_t \omega(k)\, i_{ds}(k) + \frac{1}{L_q}\Delta_t u_{qs}(k), \\
i_{ds}(k+1) &= (1 - \frac{R_s}{L_d}\Delta_t) i_{ds}(k) + \frac{n_p L_q}{L_d}\Delta_t \omega(k)\, i_{qs}(k) \\
&\quad + \frac{1}{L_d}\Delta_t u_{ds}(k),
\end{aligned}$$

where Δ_t is the sampling period. In order to simplify the mathematical model, the new variables are introduced as below:

$$\begin{aligned}
x_1(k) &= \theta(k), x_2(k) = \omega(k), x_3(k) = i_{qs}(k), \\
x_4(k) &= i_{ds}(k), a_1 = \frac{3n_p\Phi}{2J}, a_2 = \frac{3n_p(L_d - L_q)}{2J}, \\
a_3 &= \frac{B}{J}, a_4 = \frac{1}{J}, b_1 = \frac{R_s}{L_q}, \\
b_2 &= \frac{n_p\Phi}{L_q}, b_3 = \frac{n_p L_d}{L_q}, b_4 = \frac{1}{L_q}, \\
c_1 &= \frac{R_s}{L_d}, c_2 = \frac{n_p L_q}{L_d}, c_3 = \frac{1}{L_d}.
\end{aligned}$$

Based on the above substitution, the discrete-time dynamic model of IPMSM drivers can be represented by as follows:

$$\begin{aligned}
x_1(k+1) &= x_1(k) + \Delta_t x_2(k), \\
x_2(k+1) &= a_1\Delta_t x_3(k) + a_2\Delta_t x_3(k)\, x_4(k) + (1 - a_3\Delta_t)\, x_2(k) \\
&\quad - a_4\Delta_t T_L, \\
x_3(k+1) &= (1 - b_1\Delta_t)\, x_3(k) - b_2\Delta_t x_2(k) - b_3\Delta_t x_2(k)\, x_4(k) + b_4\Delta_t u_{qs}(k), \\
x_4(k+1) &= (1 - c_1\Delta_t)\, x_4(k) + c_2\Delta_t x_2(k)\, x_3(k) + c_3\Delta_t u_{ds}(k).
\end{aligned} \tag{10.1}$$

Remark 10.1 It should be pointed out that the coupling nonlinear term $a_2\Delta_t x_3(k)$ $x_4(k)$ (because of $L_d \neq L_q$) within the above model (10.1) makes the discrete-time IPMSM drive system more complex than the model of PMSM described in [22], which adds the coupling nonlinearity and complexity and will make the backstepping design difficult. ♦

The control objective is to design an adaptive fuzzy controller such that the state variable $x_1(k)$ follows the given reference signal $x_d(k)$ and all the closed-loop signals are bounded. The approximation property of the fuzzy logic systems (FLS) can be found in [23].

10.3 Discrete-Time Fuzzy Control for IPMSM

In this section, we will design an adaptive fuzzy control for the discrete-time IPMSM drive system via backstepping.

Step 1: For the reference signal x_d, define the tracking error variable as $e_1(k) = x_1(k) - x_d(k)$. From the first equation of (10.1), we can gain

$$e_1(k+1) = x_1(k+1) - x_d(k+1) = x_1(k) + \Delta_t x_2(k) - x_d(k+1).$$

The Lyapunov function candidate is chosen as $V_1(k) = \frac{1}{2}e_1^2(k)$, then the difference of $V_1(k)$ is computed by

$$\begin{aligned}\Delta V_1(k) &= \frac{1}{2}e_1^2(k+1) - \frac{1}{2}e_1^2(k)\\ &= \frac{1}{2}[x_1(k) + \Delta_t x_2(k) - x_d(k+1)]^2 - \frac{1}{2}e_1^2(k).\end{aligned} \tag{10.2}$$

The virtual control law $\alpha_1(k)$ is constructed as

$$\alpha_1(k) = \frac{1}{\Delta_t}[-x_1(k) + x_d(k+1)]. \tag{10.3}$$

By use of (10.2), (10.3) can be expressed to the following form:

$$\Delta V_1(k) = \frac{1}{2}\Delta_t^2 e_2^2(k) - \frac{1}{2}e_1^2(k), \tag{10.4}$$

with $e_2(k) = x_2(k) - \alpha_1(k)$.

Step 2: From the second equation of (10.1), we can obtain

$$\begin{aligned}e_2(k+1) &= x_2(k+1) - \alpha_1(k+1)\\ &= a_1\Delta_t x_3(k) + a_2\Delta_t x_3(k)x_4(k) - a_4\Delta_t T_L + (1 - a_3\Delta_t)x_2(k)\\ &\quad -\alpha_1(k+1).\end{aligned} \tag{10.5}$$

According to (10.3), we can get

$$\alpha_1(k+1) = \frac{1}{\Delta_t}\left[-x_1(k+1) + x_d(k+2)\right]$$
$$= \frac{1}{\Delta_t}\left[-x_1(k) - \Delta_t x_2(k) + x_d(k+2)\right]. \tag{10.6}$$

Substituting (10.6) into (10.5) leads to

$$e_2(k+1) = a_1\Delta_t x_3(k) + a_2\Delta_t x_3(k)\, x_4(k) - \frac{x_d(k+2)}{\Delta_t}$$
$$+ (2 - a_3\Delta_t)\, x_2(k) - a_4\Delta_t T_L + \frac{x_1(k)}{\Delta_t}. \tag{10.7}$$

The Lyapunov function candidate is chosen as $V_2(k) = \frac{1}{2}e_2^2(k) + V_1(k)$. Then the difference of $V_2(k)$ is given by

$$\Delta V_2(k) = \frac{1}{2}e_2^2(k+1) - \frac{1}{2}e_2^2(k) + \Delta V_1(k)$$
$$= \frac{1}{2}\left[f_1(k) + a_2\Delta_t x_3(k)\, x_4(k)\right]^2 - \frac{1}{2}e_2^2(k) + \Delta V_1(k), \tag{10.8}$$

where

$$f_1(k) = a_1\Delta_t x_3(k) + (2 - a_3\Delta_t)\, x_2(k) + \frac{1}{\Delta_t}x_1(k) - a_4\Delta_t T_L - \frac{1}{\Delta_t}x_d(k+2).$$

Construct $\alpha_2(k)$ as

$$\alpha_2(k) = \frac{1}{a_1\Delta_t}\Big[-(2 - a_3\Delta_t)\, x_2(k) - \frac{1}{\Delta_t}x_1(k) + a_4\Delta_t T_L$$
$$+ \frac{1}{\Delta_t}x_d(k+2)\Big]. \tag{10.9}$$

Using (10.4) and (10.9), the difference of $V_2(k)$ can be rewritten to the following form:

$$\Delta V_2(k) = \frac{1}{2}\left[a_1\Delta_t e_3(k) + a_2\Delta_t x_3(k)\, x_4(k)\right]^2 - \frac{1}{2}\left(1 - \Delta_t^2\right) e_2^2(k)$$
$$-\frac{1}{2}e_1^2(k), \tag{10.10}$$

with $e_3(k) = x_3(k) - \alpha_2(k)$.

Utilizing the fact that $(a_1\Delta_t e_3(k) + a_2\Delta_t x_3(k)\, x_4(k))^2 \leqslant 2a_1^2\Delta_t^2 e_3^2(k) + 2a_2^2 \Delta_t^2 x_3^2(k)\, x_4^2(k)$, we can obtain

$$\Delta V_2(k) \leqslant a_1^2 \Delta_t^2 e_3^2(k) + a_2^2 \Delta_t^2 x_3^2(k) x_4^2(k) - \frac{1}{2}\left(1 - \Delta_t^2\right) e_2^2(k) - \frac{1}{2} e_1^2(k). \tag{10.11}$$

Step 3: From the third equation of (10.1), we can obtain

$$e_3(k+1) = x_3(k+1) - \alpha_2(k+1) = f_3(k) + b_4 \Delta_t u_{qs}(k), \tag{10.12}$$

where

$$f_3(z_3(k)) = (1 - b_1 \Delta_t) x_3(k) - b_2 \Delta_t x_2(k) - b_3 \Delta_t x_2(k) x_4(k) - \alpha_2(k+1),$$

$z_3(k) = [x_1(k), x_2(k), x_3(k), x_4(k), x_d(k), x_d(k+1), x_d(k+2), x_d(k+3)]^T$ and $\alpha_2(k+1)$ can be obtained from equality (10.9).

Remark 10.2 The virtual controller $\alpha_2(k+1)$ contains future information. If we continue to construct the real controller via backstepping, we will end up with a controller containing more future information, and make it infeasible in practice. This drawback was called a noncausal problem [20]. The existing result to solve this problem is to transform the systems into a predictor form, which will add the control complexity. In this paper, we use the recursion formula to gain time k to indicate $\alpha_2(k+1)$, thus the noncausal problem can be overcome. ♦

Choose the Lyapunov function candidate as

$$V_3(k) = \frac{1}{2} e_3^2(k) + V_2(k).$$

Furthermore, differencing $V_3(k)$ yields

$$\begin{aligned} \Delta V_3(k) &= \frac{1}{2} e_3^2(k+1) - \frac{1}{2} e_3^2(k) + \Delta V_2(k) \\ &= \frac{1}{2}\left[f_3(z_3(k)) + b_4 \Delta_t u_{qs}(k)\right]^2 - \frac{1}{2} e_3^2(k) + \Delta V_2(k). \end{aligned} \tag{10.13}$$

Remark 10.3 Noting that $f_3(z_3(k))$ contains $\alpha_2(k+1)$ and the nonlinear term $b_3 \Delta_t x_2(k) x_4(k)$, which will make the backstepping design becomes very difficult, and the designed control law $u_{qs}(k)$ will have a complex structure. Hence the fuzzy logic systems are used to approximate the nonlinear function $f_3(z_3(k))$ in order to simplify the structure of the control signal. ♦

By use of the approximation property of the FLS, for any given $\varepsilon_3 > 0$, there exists a fuzzy logic system $W_3^T S_3(z_3(k))$ such that

$$f_3(z_3(k)) = W_3^T S_3(z_3(k)) + \varepsilon_3, \tag{10.14}$$

where ε_3 is the approximation error. In general, W_3 is bounded and unknown. Define $\|W_3\| = \eta_3$ where $\eta_3 > 0$ is unknown constant. Let $\hat{\eta}_3(k)$ be the estimate of η_3 and $\tilde{\eta}_3(k) = \eta_3 - \hat{\eta}_3(k)$.

Now choose the following control law $u_{qs}(k)$ and adaptive law $\hat{\eta}_3(k+1)$ as

$$u_{qs}(k) = -\frac{1}{b_4\Delta_t}\hat{\eta}_3(k)\,\|S_3(z_3(k))\|, \tag{10.15}$$

$$\hat{\eta}_3(k+1) = \hat{\eta}_3(k) + \gamma_3\|S_3(z_3(k))\|e_3(k+1) - \delta_3\hat{\eta}_3(k), \tag{10.16}$$

where γ_3 and δ_3 are the positive parameters. Furthermore, using equality (10.11), (10.14) and (10.15), (10.13) can be easily verified that

$$\begin{aligned}\Delta V_3(k) &\le \frac{1}{2}[\eta_3\|S_3(z_3(k))\| + \hat{\eta}_3(k)\,\|S_3(z_3(k))\| + \varepsilon_3]^2 - \frac{1}{2}e_3^2(k) + \Delta V_2(k)\\ &\le \frac{1}{2}[2\eta_3\|S_3(z_3(k))\| - \tilde{\eta}_3(k)\,\|S_3(z_3(k))\| + \varepsilon_3]^2 - \frac{1}{2}e_3^2(k) + \Delta V_2(k)\\ &\le 4\eta_3^2\|S_3(z_3(k))\|^2 + \tilde{\eta}_3^2(k)\,\|S_3(z_3(k))\|^2 - (\frac{1}{2} - a_1^2\Delta_t^2)e_3^2(k)\\ &\quad -\frac{1}{2}e_1^2(k) - \frac{1}{2}\left(1-\Delta_t^2\right)e_2^2(k) + \varepsilon_3^2 + a_2^2\Delta_t^2x_3^2(k)\,x_4^2(k).\end{aligned} \tag{10.17}$$

Step 4: Define the tracking error variable as $e_4(k) = x_4(k)$. From the fourth equation of (10.1), we can obtain

$$e_4(k+1) = (1 - c_1\Delta_t)\,x_4(k) + c_3\Delta_t u_{ds}(k) + c_2\Delta_t x_2(k)\,x_3(k). \tag{10.18}$$

Choose the Lyapunov function candidate as $V_4(k) = \frac{P}{2}e_4^2(k) + V_3(k)$ with $P > 0$, then the difference of $V_4(k)$ is computed by

$$\begin{aligned}\Delta V_4(k) &= \frac{P}{2}e_4^2(k+1) - \frac{P}{2}e_4^2(k) + \Delta V_3(k)\\ &= \frac{P}{2}\,[f_4(k) + c_3\Delta_t u_{ds}(k)]^2 - \frac{P}{2}e_4^2(k) + \Delta V_3(k),\end{aligned} \tag{10.19}$$

where $f_4(z_4(k)) = (1 - c_1\Delta_t)\,x_4(k) + c_2\Delta_t x_2(k)\,x_3(k)$ and $z_4(k) = [x_2(k), x_3(k), x_4(k)]^T$.

Similarly, the fuzzy logic system $W_4^T S_4(z_4(k))$ is utilized to approximate the nonlinear function $f_4(z_4(k))$ in order to simplify the controller design and

$$f_4(z_4(k)) = W_4^T S_4(z_4(k)) + \varepsilon_4, \tag{10.20}$$

where $\varepsilon_4 > 0$ is the approximation error. Similar to W_3, W_4 is also unknown and bounded. Define $\|W_4\| = \eta_4$ where $\eta_4 > 0$ is unknown constant. Let $\hat{\eta}_4(k)$ be the estimate of η_4 and $\tilde{\eta}_4(k) = \eta_4 - \hat{\eta}_4(k)$.

Now choose the following control law $u_{ds}(k)$ and adaptive law $\hat{\eta}_4(k+1)$ as

$$u_{ds}(k) = -\frac{1}{c_3 \Delta_t}\hat{\eta}_4(k)\,\|S_4(z_4(k))\|, \tag{10.21}$$

$$\hat{\eta}_4(k+1) = \hat{\eta}_4(k) + \gamma_4\|S_4(z_4(k))\|e_4(k+1) - \delta_4\hat{\eta}_4(k), \tag{10.22}$$

where γ_4 and δ_4 are the positive parameters.

Remark 10.4 Note that a similar approximation-based discrete-time controller is constructed for induction motor in [16]. From controllers (10.15) and (10.21), the number of the adaptive parameter in our proposed scheme is reduced to two only, which is much less than the number of the adaptive parameters in [16]. As a result, the computational burden of the scheme is dramatically reduced, which will render our designed scheme more suitable for practical applications. ♦

Using equalities (10.17), (10.20) and (10.21), it can be shown that

$$\begin{aligned}
\Delta V_4(k) \le\ & \frac{P}{2}[\eta_4\|S_4(z_4(k))\| + \hat{\eta}_4(k)\,\|S_4(z_4(k))\| + \varepsilon_4]^2 - \frac{P}{2}e_4^2(k) + \Delta V_3(k) \\
\le\ & \frac{P}{2}[2\eta_4\|S_4(z_4(k))\| - \tilde{\eta}_4(k)\,\|S_4(z_4(k))\| + \varepsilon_4]^2 - \frac{P}{2}e_4^2(k) + \Delta V_3(k) \\
\le\ & 4P\eta_4^2\|S_4(z_4(k))\|^2 + P\tilde{\eta}_4^2(k)\,\|S_4(z_4(k))\|^2 + 4\eta_3^2\|S_3(z_3(k))\|^2 \\
& + \tilde{\eta}_3^2(k)\,\|S_3(z_3(k))\|^2 - (\frac{1}{2} - a_1^2\Delta_t^2)e_3^2(k) - \frac{1}{2}\left(1 - \Delta_t^2\right)e_2^2(k) \\
& - [\frac{P}{2} - a_2^2\Delta_t^2 x_3^2(k)]e_4^2(k) - \frac{1}{2}e_1^2(k) + \varepsilon_3^2 + P\varepsilon_4^2.
\end{aligned} \tag{10.23}$$

Remark 10.5 It can be observed that the fuzzy-approximation-based adaptive tracking control scheme is proposed and the problems of the coupling nonlinearity because of $L_d \neq L_q$ and noncausal issue for backstepping of IPMSM drive system can be overcome without transforming the system model into a predictor form [20]. ♦

From the above analysis, we now present our first main result in this paper as follows.

Theorem 10.6 *Consider system (10.1) satisfying assumptions 1-2 and the given reference signal x_d. The proposed adaptive discrete-time controllers (10.15), (10.21) and the adaptive laws (10.16) and (10.22) can guarantee the tracking error of the closed-loop controlled system converge to a sufficiently small neighborhood of the origin and all the closed-loop signals will be bounded.*

Proof To address the stability analysis of the resulting closed-loop system, the Lyapunov function candidate is chosen as

$$V(k) = V_4(k) + \frac{1}{2\gamma_3}\tilde{\eta}_3^2(k) + \frac{P}{2\gamma_4}\tilde{\eta}_4^2(k), \tag{10.24}$$

where γ_3 and γ_4 are positive parameters. Furthermore, differencing $V(k)$ yields

$$\Delta V(k) = \Delta V_4(k) + \frac{1}{2\gamma_3}[\tilde{\eta}_3^2(k+1) - \tilde{\eta}_3^2(k)] + \frac{P}{2\gamma_4}[\tilde{\eta}_4^2(k+1) - \tilde{\eta}_4^2(k)]. \tag{10.25}$$

As defined before, we can obtain

$$\tilde{\eta}_3^2(k+1) - \tilde{\eta}_3^2(k) = \eta_3^2 + \hat{\eta}_3^2(k+1) - 2\eta_3\hat{\eta}_3(k+1) - \tilde{\eta}_3^2(k). \tag{10.26}$$

Using (10.16), we can get

$$\begin{aligned}\hat{\eta}_3^2(k+1) &= [\hat{\eta}_3(k) + \gamma_3\|S_3(z_3(k))\|e_3(k+1) - \delta_3\hat{\eta}_3(k)]^2 \\ &= (1-\delta_3)^2\hat{\eta}_3^2(k) + 2(1-\delta_3)\gamma_3\|S_3(z_3(k))\|e_3(k+1)\hat{\eta}_3(k) \\ &\quad + \gamma_3^2 e_3^2(k+1)\|S_3(z_3(k))\|^2,\end{aligned} \tag{10.27}$$

$$\begin{aligned}\eta_3\hat{\eta}_3(k+1) &= \eta_3[\hat{\eta}_3(k) + \gamma_3\|S_3(z_3(k))\|e_3(k+1) - \delta_3\hat{\eta}_3(k)] \\ &= (1-\delta_3)\eta_3\hat{\eta}_3(k) + \gamma_3\|S_3(z_3(k))\|e_3(k+1)\eta_3.\end{aligned} \tag{10.28}$$

Substituting (10.27) and (10.28) into (10.26) gives us

$$\begin{aligned}&\tilde{\eta}_3^2(k+1) - \tilde{\eta}_3^2(k) = \eta_3^2 + (1-\delta_3)^2\hat{\eta}_3^2(k) + \gamma_3^2 e_3^2(k+1)\|S_3(z_3(k))\|^2 \\ &-2(1-\delta_3)\eta_3\hat{\eta}_3(k) + 2(1-\delta_3)\gamma_3\|S_3(z_3(k))\|e_3(k+1)\hat{\eta}_3(k) \\ &-\tilde{\eta}_3^2(k) - 2\gamma_3\|S_3(z_3(k))\|e_3(k+1)\eta_3.\end{aligned} \tag{10.29}$$

Then, using $\|S_3(z_3(k))\|^2 \leq 1$ and according to Young's inequality, we have

$$\begin{aligned}2\gamma_3\|S_3(z_3(k))\|e_3(k+1)\hat{\eta}_3(k) &\leq \gamma_3^2 e_3^2(k+1) + \hat{\eta}_3^2(k), \\ -2\|S_3(z_3(k))\|e_3(k+1)\eta_3 &\leq e_3^2(k+1) + \eta_3^2, \\ \gamma_3^2 e_3^2(k+1)\|S_3(z_3(k))\|^2 &\leq \gamma_3^2 e_3^2(k+1), \\ -2\eta_3\hat{\eta}_3(k) &\leq \eta_3^2 + \hat{\eta}_3^2(k).\end{aligned} \tag{10.30}$$

Substituting (10.14), (10.15) into (10.12) leads to

$$e_3(k+1) = W_3^T S_3(z_3(k)) + \varepsilon_3 + b_4\Delta_t u_{qs}(k).$$

Then, we can obtain

$$
\begin{aligned}
e_3^2(k+1) &\le [\eta_3\|S_3(z_3(k))\| + \hat{\eta}_3(k)\|S_3(z_3(k))\| + \varepsilon_3]^2 \\
&\le [2\eta_3\|S_3(z_3(k))\| - \tilde{\eta}_3(k)\|S_3(z_3(k))\| + \varepsilon_3]^2 \\
&\le \left[2\eta_3 - \tilde{\eta}_3(k) + \varepsilon_3\right]^2 \\
&\le 8\eta_3^2 + 2\tilde{\eta}_3^2(k) + 2\varepsilon_3^2.
\end{aligned} \tag{10.31}
$$

Substituting (10.30) and (10.31) into (10.29) yields

$$
\begin{aligned}
&\tilde{\eta}_3^2(k+1) - \tilde{\eta}_3^2(k) \le (16\gamma_3^2 - 8\gamma_3^2\delta_3 + 9\gamma_3 - \delta_3 + 2)\eta_3^2 \\
&+(\delta_3^2 - 4\delta_3 + 3)\hat{\eta}_3^2(k) + (4\gamma_3^2 - 2\gamma_3^2\delta_3 + 2\gamma_3 - 1)\tilde{\eta}_3^2(k) \\
&+(4\gamma_3^2 - 2\gamma_3^2\delta_3 + 2\gamma_3)\varepsilon_3^2.
\end{aligned} \tag{10.32}
$$

Similarly, we can get

$$
\begin{aligned}
&\tilde{\eta}_4^2(k+1) - \tilde{\eta}_4^2(k) \le (16\gamma_4^2 - 8\gamma_4^2\delta_4 + 9\gamma_4 - \delta_4 + 2)\eta_4^2 \\
&+(\delta_4^2 - 4\delta_4 + 3)\hat{\eta}_4^2(k) + (4\gamma_4^2 - 2\gamma_4^2\delta_4 + 2\gamma_4 - 1)\tilde{\eta}_4^2(k) \\
&+(4\gamma_4^2 - 2\gamma_4^2\delta_4 + 2\gamma_4)\varepsilon_4^2.
\end{aligned} \tag{10.33}
$$

Then substituting (10.23), (10.32) and (10.33) into (10.25), we have

$$
\begin{aligned}
\Delta V(k) \le& -[\frac{P}{2} - a_2^2\Delta_t^2 x_3^2(k)]e_4^2(k) - (\frac{1}{2} - a_1^2\Delta_t^2)e_3^2(k) \\
&-\frac{1}{2}\left(1 - \Delta_t^2\right)e_2^2(k) - \frac{1}{2}e_1^2(k) + \frac{1}{2\gamma_3}[(\delta_3^2 \\
&-4\delta_3 + 3)\hat{\eta}_3^2(k) + \beta_3 + (4\gamma_3^2 - 2\gamma_3^2\delta_3 \\
&+4\gamma_3 - 1)\tilde{\eta}_3^2(k)] + \frac{P}{2\gamma_4}[(\delta_4^2 - 4\delta_4 + 3)\hat{\eta}_4^2(k) \\
&+\beta_4 + (4\gamma_4^2 - 2\gamma_4^2\delta_4 + 4\gamma_4 - 1)\tilde{\eta}_4^2(k)],
\end{aligned}
$$

where $\beta_3 = (4\gamma_3^2 - 2\gamma_3^2\delta_3 + 4\gamma_3)\varepsilon_3^2 + (16\gamma_3^2 - 8\gamma_3^2\delta_3 + 17\gamma_3 - \delta_3 + 2)\eta_3^2$ and $\beta_4 = (4\gamma_4^2 - 2\gamma_4^2\delta_4 + 4\gamma_4)\varepsilon_4^2 + (16\gamma_4^2 - 8\gamma_4^2\delta_4 + 17\gamma_4 - \delta_4 + 2)\eta_4^2$ are bounded. Define $x_3^2(k) \leqslant M$, where M is positive constant. Furthermore,

$$
\begin{aligned}
\Delta V(k) \le& -[\frac{P}{2} - a_2^2\Delta_t^2 M]e_4^2(k) - (\frac{1}{2} - a_1^2\Delta_t^2)e_3^2(k) \\
&-\frac{1}{2}\left(1 - \Delta_t^2\right)e_2^2(k) - \frac{1}{2}e_1^2(k) + \frac{1}{2\gamma_3}[(\delta_3^2 \\
&-4\delta_3 + 3)\hat{\eta}_3^2(k) + \beta_3 + (4\gamma_3^2 - 2\gamma_3^2\delta_3 \\
&+4\gamma_3 - 1)\tilde{\eta}_3^2(k)] + \frac{P}{2\gamma_4}[(\delta_4^2 - 4\delta_4 + 3)\hat{\eta}_4^2(k) \\
&+\beta_4 + (4\gamma_4^2 - 2\gamma_4^2\delta_4 + 4\gamma_4 - 1)\tilde{\eta}_4^2(k)].
\end{aligned}
$$

By choosing a suitable parameter P and sampling period Δ_t, we can get $\frac{P}{2} - a_2^2\Delta_t^2 M > 0$, $\frac{1}{2} - a_1^2\Delta_t^2 > 0$ and $1 - \Delta_t^2 > 0$. If we choose the design parameters as follows:

$$\delta_i^2 - 4\delta_i + 3 < 0, 4\gamma_i^2 - 2\gamma_i^2\delta_i + 4\gamma_i - 1 < 0,$$

for $i = 3, 4$. Then

$$\Delta V(k) \leqslant 0,$$

once the error $|e_3(k)| > \sqrt{\frac{\beta_3}{\gamma_3(1-2a_1^2\Delta_t^2)}}$ and $|e_4(k)| > \sqrt{\frac{P\beta_4}{\gamma_4(P-2a_2^2\Delta_t^2 M)}}$. That implies that the tracking error $e_i(k)(i = 1, 2, 3, 4)$ are bounded in a compact set [20].

Subtracting η_3 from both sides of (10.16), we can obtain

$$-\tilde{\eta}_3(k+1) = -\tilde{\eta}_3(k) + \gamma_3\|S_3(z_3(k))\|e_3(k+1) - \delta_3\hat{\eta}_3(k).$$

Noting that

$$\eta_3 = \tilde{\eta}_3(k) + \hat{\eta}_3(k),$$

then,

$$\tilde{\eta}_3(k+1) = (1-\delta_3)\tilde{\eta}_3(k) - \gamma_3\|S_3(z_3(k))\|e_3(k+1) + \delta_3\eta_3.$$

Choose a suitable δ_3 and let $0 < 1 - \delta_3 < 1$. Noting $||S_3(z_3(k))||$, $e_3(k)$, $\delta_3\eta_3$ are bounded and according to Lemma 1 in [24], $\tilde{\eta}_3(k)$ must be bounded in a compact set. Similarly, $\tilde{\eta}_4(k)$ must be bounded in a compact set. So, the boundedness of $\tilde{\eta}_3(k)$ and $\tilde{\eta}_4(k)$ are obtained. Then the input u_{qs} and u_{ds} are bounded. This can guarantee that all the signals including $e_i(k)(i = 1, 2, 3, 4)$, $\hat{\eta}_3(k)$, $\hat{\eta}_4(k)$ are bounded and $\lim_{k\to\infty}\|x_1(k) - x_d(k)\| \leqslant \sigma$ where σ is small positive constant. ■

10.4 Simulation Results

To illustrate the effectiveness of the proposed results, the simulation is run for IPMSM with the parameters [13]:

$$J = 0.00379\,\text{Kgm}^2,\ R_s = 0.68,\ L_d = 0.00315\text{H},$$
$$n_p = 3,\ L_q = 0.00285\text{H},\ \Phi = 0.1245\text{H},$$
$$B = 0.001158\,\text{Nm/(rad/s)}.$$

The control objective is to design a controller such that $x_1(k)$ tracks the reference signal $x_d(k)$ effectively. The reference signal is chosen as $x_d(k) = 2\cos(\Delta_t k\pi/2)$.

The load torque disturbances are introduced to assess the motor recovery ability under our proposed controllers and the load torque is given as follows.

$$T_L = \begin{cases} 1.5, 0 \leq k \leq 2000, \\ 3, k \geq 2000. \end{cases}$$

The initial values of the states are chosen as $x_1(0) = x_2(0) = x_3(0) = x_4(0) = 0$.The sampling period is chosen as $\Delta_t = 0.0055$s. The values of the design parameters were selected as $\delta_3 = 0.39$, $\delta_4 = 0.29$, $\gamma_3 = 0.3$ and $\gamma_4 = 0.7$.

Remark 10.7 For the discrete-time control system, the selection of sampling period Δ_t is a critical issue. If the sampling period Δ_t is too large, the sample accuracy would be poor and will bring down the control system performance. Decreasing Δ_t will gain a more precisely discrete-time dynamic model of IPMSMs, but it will add system control burden such as the computation burden. Therefore, according to the control performance and system control burden, we choose a suitable value of Δ_t in this paper. ♦

The simulation results are described in Figs. 10.1, 10.2, 10.3, 10.4, 10.5, 10.6, 10.7 and 10.8. Figure 10.1 shows the trajectories of $x_1(k)$ and $x_d(k)$, where the solid line represents $x_1(k)$ and the dashed line represents $x_d(k)$. The dynamics of the tracking error is shown in Fig. 10.2. It can be observed that the controllers copes easily with the sudden change on the load torque and provides a fast position

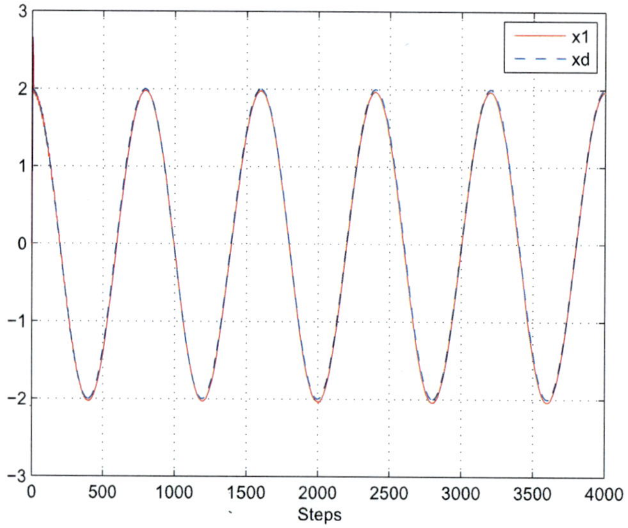

Fig. 10.1 Trajectories of the x_1 and x_{1d}

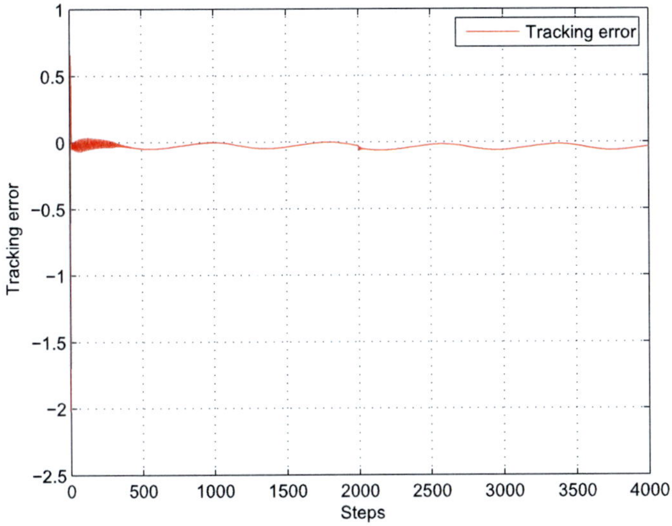

Fig. 10.2 The tracking error of x_1 and x_{1d}

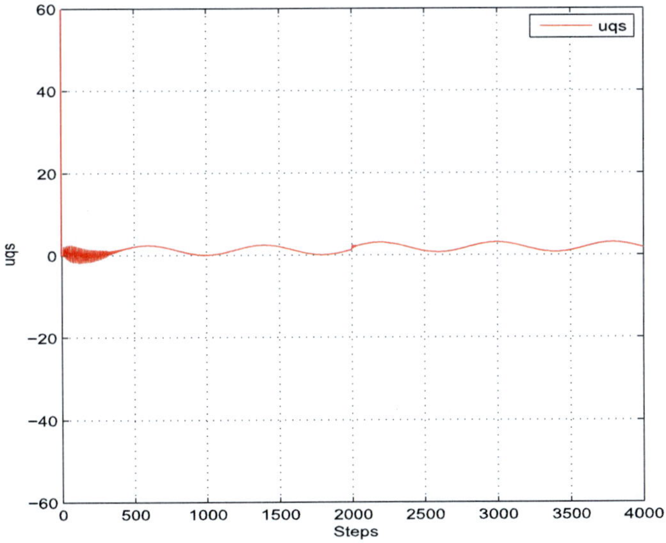

Fig. 10.3 Curve of the u_{qs}

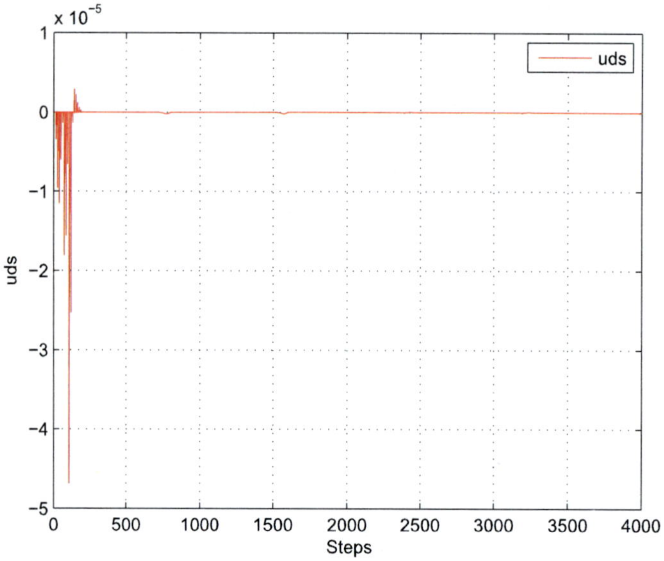

Fig. 10.4 Curve of the u_{ds}

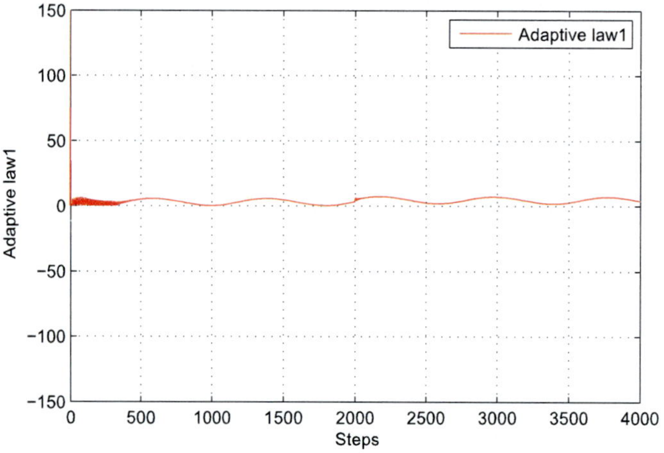

Fig. 10.5 Curve of the adaptive law $\hat{\eta}_3\ (k)$

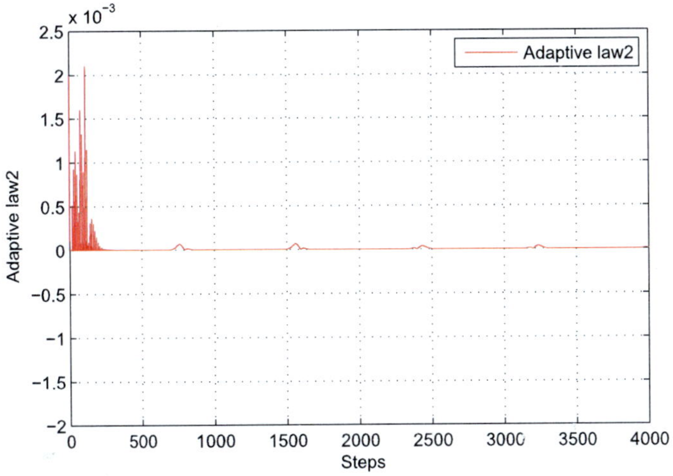

Fig. 10.6 Curve of the adaptive law $\hat{\eta}_4(k)$

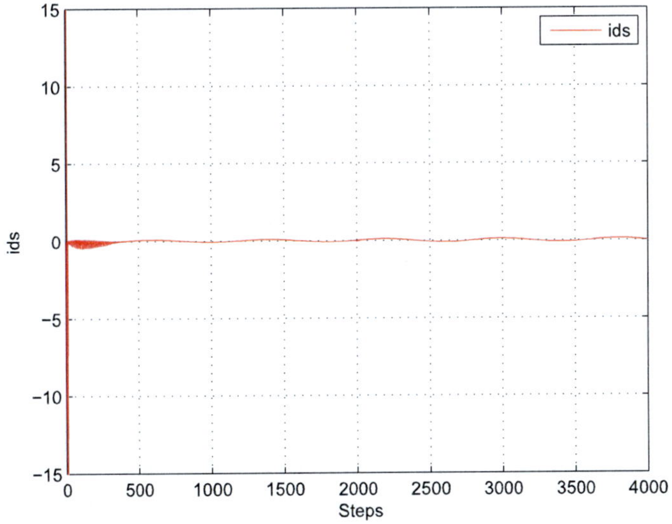

Fig. 10.7 Curve of the i_{qs}

tracking response when the load torque changes. Moreover, the position tracking error remains small and without overshoot, which produces smooth current signals. Figures 10.3 and 10.4 shows the trajectories of $u_{qs}(k)$ and $u_{ds}(k)$. Furthermore, the system adaptive laws are given in Figs. 10.5 and 10.6 to demonstrate the adaptive learning performance. Boundedness of $i_{qs}(k)$ and $i_{ds}(k)$ are illustrated by Figs. 10.7 and 10.8. From Figs. 10.3, 10.4 and Figs. 10.7, 10.8, it can be seen that boundedness

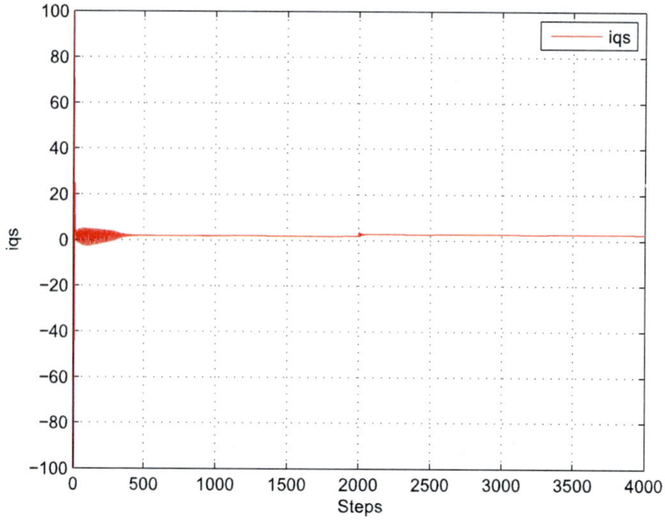

Fig. 10.8 Curve of the i_{ds}

of $u_{ds}(k)$, $u_{qs}(k)$, $i_{qs}(k)$ and $i_{ds}(k)$ are verified. The controllers can guarantee the robustness against the system parameter variations and load disturbances.

10.5 Conclusion

In this chapter, a discrete-time adaptive position tracking control method was proposed for IPMSM based on fuzzy-approximation. The accurate approximate discrete-time IPMSM position tracking system model is derived by the Euler direct discretization method. Fuzzy logic systems are used to approximate the nonlinearities of the discrete-time IPMSM drive system. The proposed discrete-time fuzzy controller can guarantee the tracking error converges to a small neighborhood of the origin and all the signals are bounded. Simulation results are given to demonstrate the effectiveness and the potentials of the theoretic results obtained.

References

1. Duan, Z., Ghous, I., Shen, J.: Fault detection observer design for discrete-time 2-D T-S fuzzy systems with finite-frequency specifications. Fuzzy Sets Syst. **392**(1), 24–45 (2020)
2. Adzhiev, S.Z., Melikhov, I.V., Vedenyapin, V.V.: On the H-theorem for the Becker-Doring system of equations for the cases of continuum approximation and discrete time. Physica A **553**(1), 124608 (2020)

3. Deplano, D., Franceschelli, M., Giua, A.: A nonlinear Perron-Frobenius approach for stability and consensus of discrete-time multi-agent systems. Automatica **118**, 109025 (2020)
4. Huang, M., Liu, C., He, X., Ma, L., Lu, Z., Su, H.: Reinforcement learning-based control for nonlinear discrete-time systems with unknown control directions and control constraints. Neurocomputing **402**(18), 50–65 (2020)
5. Treesatayapun, C.: Robotic architecture as unknown discrete-time system based on variable-frequency drive and adaptive controller. Robot. Comput.-Integr. Manuf. **64**, 101951 (2020)
6. Zhang, Z., Liang, H., Wu, C., Ahn, C.K.: Adaptive event-triggered output feedback fuzzy control for nonlinear networked systems with packet dropouts and actuator failure. IEEE Trans. Fuzzy Syst. **27**(9), 1793–1806 (2019)
7. Shen, H., Li, F., Wu, Z.-G., Park, J.H., Sreeram, V.: Fuzzy-model-based nonfragile control for nonlinear singularly perturbed systems with semi-Markov jump parameters. IEEE Trans. Fuzzy Syst. **26**(6), 3428–3439 (2018)
8. Sun, X., Chen, L., Jiang, H., Yang, Z., Chen, J., Zhang, W.: High-performance control for a bearingless permanent-magnet synchronous motor using neural network inverse scheme plus internal model controllers. IEEE Trans. Ind. Electron. **63**(6), 3479–3488 (2016)
9. Sun, X., Shi, Z., Chen, L., Yang, Z.: Internal model control for a bearingless permanent magnet synchronous motor based on inverse system method. IEEE Trans. Energy Convers. **31**(4), 1539–1548 (2016)
10. Zhang, X., Hou, B., Mei, Y.: Deadbeat predictive current control of permanent-magnet synchronous motors with stator current and disturbance observer. IEEE Trans. Power Electron. **32**(5), 3818–3834 (2017)
11. Li, S.H., Liu, Z.G.: Adaptive speed control for permanent-magnet synchronous motor system with variations of load inertia. IEEE Trans. Ind. Electron. **56**(8), 3050–3059 (2009)
12. Li, S.H., Gu, H.: Fuzzy adaptive internal model control schemes for PMSM speed-regulation system. IEEE Trans. Ind. Inform. **8**(4), 767–779 (2012)
13. Yu, J.P., Ma, Y.M., Chen, B., Yu, H.S.: Adaptive fuzzy backstepping position tracking control for a permanent magnet synchronous motor. Int. J. Innov. Comput., Inf. Control **7**(4), 1589–1602 (2011)
14. Yu, J.P., Chen, B., Yu, H.S., Gao, J.W.: Adaptive fuzzy tracking control for the chaotic permanent magnet synchronous motor drive system via backstepping. Nonlinear Anal. RWA **12**(1), 671–681 (2011)
15. Zhang, H.B., Dang, C.Y., Li, C.G.: Decentralized $H\infty$ filter design for discrete-time interconnected fuzzy systems. IEEE Trans. Fuzzy Syst. **17**(6), 1428–1440 (2009)
16. Alanis, A.Y., Sanchez, E.N., Loukianov, A.G.: Real-time discrete backstepping neural control for induction motors. IEEE Trans. Contr. Syst. Technol. **19**(2), 359–366 (2011)
17. Castaneda, C.E., Loukianov, A.G., Sanchez, E.N., Castillo-Toledo, B.: Discrete-time neural sliding-mode block control for a DC motor with controlled flux. IEEE Trans. Ind. Electron. **59**(2), 1194–1207 (2012)
18. Nesic, D., Teel, A.R.: Stabilization of sampled-data nonlinear systems via backstepping on their Euler approximate model. Automatica **42**, 1801–1808 (2006)
19. Veselic, B., Perunicic-Drazenovic, B., Milosavljevic, C.: High-performance position control of induction motor using discrete-time sliding-mode control. IEEE Trans. Ind. Electron. **55**(11), 3809–3817 (2008)
20. Liu, Y.J., Chen, C.L.P., Wen, G.C., Tong, S.C.: Adaptive neural output feedback tracking control for a class of uncertain discrete-time nonlinear systems. IEEE Trans. Neural Netw. **22**(7), 1162–1167 (2011)
21. Pillay, P., Krishnan, R.: Modeling of permanent magnet motor drives. IEEE Trans. Ind. Electron. **35**(4), 537–541 (1998)
22. Choi, H.H., Jung, J.W.: Discrete-time fuzzy speed regulator design for PM synchronous motor. IEEE Trans. Ind. Electron. **60**(2), 600–607 (2013)
23. Jagannathan, S.: Adaptive fuzzy logic control of feedback linearizable discrete-time dynamical systems under persistence of excitation. Automatica **34**(311), 1295–1310 (1998)
24. Chen, W.S.: Adaptive NN control for discrete-time pure-feedback systems with unknown control direction under amplitude and rate actuator constraints. ISA Trans. **48**(3), 304–311 (2009)

Chapter 11
Adaptive Fuzzy Tracking Control for the Chaotic PMSM Drive System

An adaptive fuzzy tracking control scheme is proposed to deal with chaos in the permanent magnet synchronous motor drive system. The fuzzy logic systems are employed to approximate unknown nonlinearities and the adaptive backstepping technique is used to construct controller. Compared with the traditional backstepping control, the structure of the designed fuzzy controller is simple. The simulation results show that the proposed control scheme can effectively suppress PMSM drive system chaos and the track the reference signal successfully when the parameters are uncertain.

11.1 Introduction

In the past two decades, chaos control research [1–5] has attracted much attention due to its important theoretical and practical value. The occurrence of chaos in motor drive systems was addressed by Kuroe and Hayashi [6] in the late 1980s. Since then, it has been one of the hottest research in nonlinear sciences. Many researchers have paid attention to the discovery of chaos and its control in several types of motor drive systems, such as DC motor drives [7–9], step motors [10], induction motor drives [11], synchronous reluctance motor drives [12, 13], switched reluctance motor drives [14] and so on [15–17].

In this chapter, an adaptive fuzzy tracking control scheme is proposed to deal with chaos in the permanent magnet synchronous motor drive systems via backstepping technology. During the controller design process, fuzzy logic systems are introduced to approximate the nonlinearities of the chaotic PMSM drive system, this means that the undeterministic parameters are taken into account, no regression matrices need to be found and the problem of "explosion of terms" is overcome. In addition, the proposed controller guarantees that the tracking error converges to a small neighborhood of the origin and all the closed-loop signals are bounded. The simulation results

J. Yu et al., *Intelligent Backstepping Control for the Alternating-Current Drive Systems*, Studies in Systems, Decision and Control 349,
https://doi.org/10.1007/978-3-030-67723-7_11

are supplied to demonstrate the effectiveness and robustness against the parameter uncertainties in the chaotic PMSM drive system.

11.2 Mathematical Model of Chaotic PMSM Drive System and Preliminaries

The dimensionless mathematical model of PMSM with the smooth air gap can be described as follows [15]:

$$\begin{aligned}\frac{d\omega}{dt} &= \sigma(i_q - \omega) - \tilde{T}_L,\\ \frac{di_q}{dt} &= -i_q - i_d\omega + \gamma\omega + \tilde{u}_q,\\ \frac{di_d}{dt} &= -i_d + i_q\omega + \tilde{u}_d,\end{aligned} \tag{11.1}$$

where ω, i_d and i_q are state variables, which denote angle speed and the $d - q$ axis currents, respectively. σ and γ are system operating parameters, which are positive. $\tilde{T}_L$, $\tilde{u}_d$ and $\tilde{u}_q$ stand for the $d - q$ axis voltages and load torque, respectively.

In system (11.1), the external inputs are set to zero, namely, $\tilde{T}_L = \tilde{u}_d = \tilde{u}_q = 0$ [15]. Then, the system (11.1) becomes an unforced system:

$$\begin{aligned}\frac{d\omega}{dt} &= \sigma(i_q - \omega),\\ \frac{di_q}{dt} &= -i_q - i_d\omega + \gamma\omega,\\ \frac{di_d}{dt} &= -i_d + i_q\omega.\end{aligned}$$

The modern nonlinear theories such as bifurcation and chaos, have been widely applied to research the PMSM driver system's stability. The research found the PMSM is experiencing chaotic behavior when the operating parameters σ and γ fall into certain areas. For example, the PMSM displays chaos with σ = 5.45 and γ = 20. The typical chaotic attractor is displayed in Fig. 11.1. These chaotic oscillations can destroy the stabilization of the PMSM drive system. In order to eliminate or control chaos, we use u_d as the manipulated variable which is desirable for the real application. Then, an adaptive fuzzy tracking control approach is developed to control chaos in the PMSM drive system via the backstepping technique. In order to facilitate calculation, the following notations are introduced: $x_1 = \omega$, $x_2 = i_q$, $x_3 = i_d$. By utilizing these notations, the dynamic mathematical model of the PMSM driver system can be represented by the following differential equations:

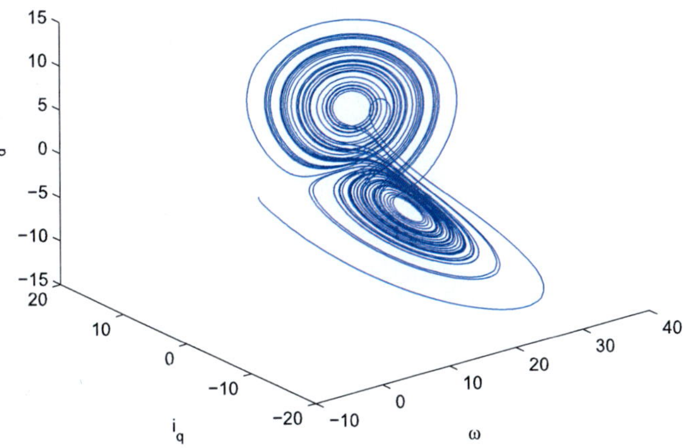

Fig. 11.1 Curves of the typical chaotic attractor in PMSM with system parameters

$$\begin{aligned}\dot{x}_1 &= \sigma(x_2 - x_1),\\ \dot{x}_2 &= -x_2 - x_1x_3 + \gamma x_1,\\ \dot{x}_3 &= -x_3 + x_1x_2 + u_d.\end{aligned} \tag{11.2}$$

The control objective is to design an adaptive fuzzy tracking controller such that the state variable x_1 follows the given reference signal x_d and all the closed-loop signals are bounded.

Lemma 11.1 *[18] Let $f(x)$ be a continuous function defined on a compact set Ω. Then for any scalar $\varepsilon > 0$, there exists a fuzzy logic system $W^T S(x)$ such that*

$$\sup_{x\in\Omega} \left| f(x) - W^T S(x) \right| \le \varepsilon,$$

where $W = [W_1, ..., W_N]^T$ is the ideal constant weight vector, and $S(x) = [p_1(x), p_2(x), ..., p_N(x)]^T / \sum_{i=1}^{N} p_i(x)$ is the basis function vector, with $N > 1$ being the number of the fuzzy rules and p_i are chosen as Gaussian functions, i.e., for $i = 1, 2, ..., N$, $p_i(x) = \exp[\frac{-(x-\mu_i)^T(x-\mu_i)}{\eta_i^2}]$ where $\mu_i = [\mu_{i1}, \mu_{i2}, ..., \mu_{in}]^T$ is the center vector, and η_i is the width of the Gaussian function.

11.3 Adaptive Fuzzy Controller with the Backstepping Technique

This section will propose an adaptive fuzzy tracking control approach to control chaos in the PMSM drive system via the backstepping. The backstepping design procedure contains 3 steps. At each design step, a virtual control function α_i $(i = 1, 2)$ will be constructed by applying an appropriate Lyapunov function. At the last step, the real controller is designed to control the system.

Step 1: For the reference signal x_d, define the tracking error variable as $z_1 = x_1 - x_d$, where x_d is a constant. From the first differential equation of (11.2), the error dynamic system is given by $\dot{z}_1 = \sigma(x_2 - x_1) - \dot{x}_d = \sigma(x_2 - x_1)$.

Select Lyapunov function candidate as $V_1 = \frac{1}{2}z_1^2$, then the time derivative of V_1 is computed by

$$\dot{V}_1 = z_1\dot{z}_1 = z_1\sigma(x_2 - x_1). \tag{11.3}$$

Construct the virtual control law α_1 as

$$\alpha_1(x_1, x_d) = -\bar{k}_1 z_1 + x_1, \tag{11.4}$$

with $\bar{k}_1$ is a positive constant. By using (11.4) and substituting $z_2 + \alpha_1$ for x_2 in (11.3), (11.3) can be rewritten in the following form.

$$\dot{V}_1 = -\bar{k}_1\sigma z_1^2 + \sigma z_1 z_2 = -k_1 z_1^2 + \sigma z_1 z_2,$$

with $k_1 > 0$ being a design parameter and $z_2 = x_2 - \alpha_1$.

Step 2: Differentiating z_2 gives

$$\dot{z}_2 = \dot{x}_2 - \dot{\alpha}_1 = -x_2 - x_1x_3 + \gamma x_1 - \dot{\alpha}_1. \tag{11.5}$$

Now, choose the Lyapunov function candidate as $V_2 = V_1 + \frac{1}{2}z_2^2$. Obviously, the time derivative of V_2 is given by

$$\dot{V}_2 = \dot{V}_1 + z_2\dot{z}_2 = -k_1 z_1^2 + z_2(\sigma z_1 - x_2 - x_1x_3 + \gamma x_1 - \dot{\alpha}_1). \tag{11.6}$$

In the realistic model of PMSM, limited to the work conditions, the parameter γ is unknown. So it cannot be used to construct the control signal. Thus, let $\hat{\gamma}$ be the estimation of γ. The corresponding adaptation laws will be specified later. The virtual control α_2 is constructed as

$$\alpha_2(Z_2) = -\frac{1}{x_1}(-k_2 z_2 - \sigma z_1 + x_2 - \hat{\gamma}x_1 + \dot{\alpha}_1), \tag{11.7}$$

where $k_2 > 0$ is a positive design parameter and $Z_2 = [x_1, x_2, x_d, \dot{x}_d, \ddot{x}_d, \hat{\gamma}]^T$. Adding and subtracting α_2 in the bracket in (11.6) shows that

$$\dot{V}_2 = -k_1 z_1^2 - k_2 z_2^2 - x_1 z_2 z_3 - z_2(\hat{\gamma} - \gamma)x_1, \tag{11.8}$$

with $z_3 = x_3 - \alpha_2$.

Step 3: Differentiating z_3 results in the following differential equation

$$\dot{z}_3 = \dot{x}_3 - \dot{\alpha}_2 = -x_3 + x_1 x_2 + u_d - \dot{\alpha}_2.$$

Select the Lyapunov function candidate as $V_3 = V_2 + \frac{1}{2} z_3^2$. Furthermore, differentiating V_3 yields

$$\begin{aligned} \dot{V}_3 &= \dot{V}_2 + z_3 \dot{z}_3 = \dot{V}_2 + z_3(-x_3 + x_1 x_2 + u_d - \dot{\alpha}_2) \\ &= -k_1 z_1^2 - k_2 z_2^2 - z_2(\hat{\gamma} - \gamma) x_1 + z_3(f_3 + u_d), \end{aligned} \tag{11.9}$$

where

$$\begin{aligned} \dot{\alpha}_1 &= \dot{x}_1 - \dot{x}_d = \sigma(x_2 - x_1) - \dot{x}_d, \\ \dot{\alpha}_2 &= \sum_{i=1}^{2} \frac{\partial \alpha_2}{\partial x_i} \dot{x}_i + \sum_{i=0}^{2} \frac{\partial \alpha_2}{\partial x_d^{(i)}} x_d^{(i+1)} + \frac{\partial \alpha_2}{\partial \hat{\gamma}} \dot{\hat{\gamma}} \\ &= \frac{\partial \alpha_2}{\partial x_1} \sigma(x_2 - x_1) + \frac{\partial \alpha_2}{\partial x_2} (-x_2 - x_1 x_3 + \gamma x_1) \\ &\quad + \sum_{i=0}^{2} \frac{\partial \alpha_2}{\partial x_d^{(i)}} x_d^{(i+1)} + \frac{\partial \alpha_2}{\partial \hat{\gamma}} \dot{\hat{\gamma}}, \\ f_3(Z) &= -x_3 + x_1 x_2 - x_1 z_2 - \dot{\alpha}_2, \\ Z &= [x_1, x_2, x_3, x_d, \dot{x}_d, \ddot{x}_d, \hat{\gamma}]^T. \end{aligned} \tag{11.10}$$

Notice that f_3 contains the derivative of α_2. This will make the classical adaptive backstepping design become very complex and troubling, and the designed control law u_d will have a complex structure. To avoid this trouble and simplify the control signal structure, we will employ the fuzzy logic system to approximate the nonlinear function f_3. As shown later, the design procedure of u_d becomes simple and u_d is of a simple structure.

According to Lemma 11.1, for any given $\varepsilon_3 > 0$, there exists a fuzzy logic system $W_3^T S(Z)$ such that

$$f_3(Z) = W_3^T S(Z) + \delta_3(Z), \tag{11.11}$$

where $\delta_3(Z)$ is the approximation error and satisfies $|\delta_3| \le \varepsilon_3$. Consequently, a straightforward calculation produces the following inequality.

$$\begin{aligned} z_3 f_3(Z) &= z_3 \left(W_3^T S(Z) + \delta_3(Z) \right) \\ &\le z_3 \left(\frac{\| W_3^T \| \, S(Z) W_3^T l_3}{l_3 \, \| W_3^T \|} + \varepsilon_3 \right) \\ &\le \frac{1}{2 l_3^2} z_3^2 \, \| W_3 \|^2 \, S^T S + \frac{1}{2} l_3^2 + \frac{1}{2} z_3^2 + \frac{1}{2} \varepsilon_3^2, \end{aligned} \tag{11.12}$$

where l_3 is a positive constant. Thus, it follows immediately from substituting (11.7) into (11.9) that

$$\dot{V}_3 \leq -k_1 z_1^2 - k_2 z_2^2 - z_2(\hat{\gamma} - \gamma)x_1 + \frac{1}{2l_3^2} z_3^2 \left\| W_3 \right\|^2 S^2 + \frac{1}{2} l_3^2 + \frac{1}{2} z_3^2 + \frac{1}{2} \varepsilon_3^2 + z_3 u_d.$$

At this present stage, the control law u_d is designed as

$$u_d = -k_3 z_3 - \frac{1}{2} z_3 - \frac{1}{2l_3^2} z_3 \hat{\theta} S^2, \tag{11.13}$$

where $\hat{\theta}$ is the estimation of the unknown constant θ which will be specified later. Define $\theta = \left\| W_3 \right\|^2$. Furthermore, using the equality (11.13), it can be verified easily that

$$\dot{V}_3 \leq -\sum_{i=1}^{3} k_i z_i^2 + \frac{1}{2l_3^2} z_3^2 (\left\| W_3 \right\|^2 - \hat{\theta}) S^T(Z) S(Z) + \frac{1}{2} l_3^2 + \frac{1}{2} \varepsilon_3^2 + z_2(\gamma - \hat{\gamma})x_1. \tag{11.14}$$

Introduce variables $\tilde{\gamma}$ and $\tilde{\theta}$ as

$$\begin{aligned} \tilde{\gamma} &= \hat{\gamma} - \gamma, \\ \tilde{\theta} &= \hat{\theta} - \theta, \end{aligned} \tag{11.15}$$

and definite the Lyapunov function candidate as

$$V = V_3 + \frac{1}{2r_1} \tilde{\gamma}^2 + \frac{1}{2r_2} \tilde{\theta}^2, \tag{11.16}$$

where r_i, $i = 1, 2$ are positive constant. By differentiating V and taking (11.14)–(11.16) into account, one can obtain

$$\begin{aligned} \dot{V} \leq & -\sum_{i=1}^{3} k_i z_i^2 + \frac{1}{2l_3^2} z_3^2 \tilde{\theta} S^T(Z) S(Z) + z_2(\gamma - \hat{\gamma})x_1 \\ & + \frac{1}{2} l_3^2 + \frac{1}{2} \varepsilon_3^2 + \frac{1}{r_1} \tilde{\gamma} \dot{\hat{\gamma}} + \frac{1}{r_2} \tilde{\theta} \dot{\hat{\theta}} \\ = & -\sum_{i=1}^{3} k_i z_i^2 + \frac{1}{2} l_3^2 + \frac{1}{2} \varepsilon_3^2 + \frac{1}{r_1} \tilde{\gamma} \left(-r_1 z_2 x_1 + \dot{\hat{\gamma}} \right) \\ & + \frac{1}{r_2} \tilde{\theta} \left[-\frac{1}{2l_3^2} z_3^2 S^T(Z) S(Z) + \dot{\hat{\theta}} \right]. \end{aligned} \tag{11.17}$$

According to (11.17), the corresponding adaptive laws are chosen as follows:

$$\begin{aligned} \dot{\hat{\gamma}} &= r_1 z_2 \dot{\alpha}_1 - m_1 \hat{\gamma}, \\ \dot{\hat{\theta}} &= \frac{1}{2l_3^2} z_3^2 S^T(Z) S(Z) - m_2 \hat{\theta}, \end{aligned} \tag{11.18}$$

where m_i, for $i = 1, 2$ are positive constants.

Remark 11.2 To demonstrate the advantage of the adaptive fuzzy backstepping technique over the conventional backstepping summarized, we compare the controller in Eqs. (11.4), (11.7) and (11.13) with those described in Eqs. (11.24), (11.27) and (11.30) corresponding, respectively. It can be seen clearly that the backstepping controller (11.30) would be much more complicated than that of the new controller (11.13). The number of terms in the expression of (11.30) is much larger. This drawback is called the "explosion of terms" above [19]. ♦

Remark 11.3 In the realistic model of PMSM, the system parameters σ and γ may be unknown, so, they cannot be used to construct the control signal unless we specify its corresponding adaptation law. Since the undeterministic parameter σ does not influence the final control law, we only introduce the adaptive parameter scalar, i.e., $\hat{\gamma}$. ♦

Theorem 11.4 *Consider the system (11.2) and the reference signal x_d. Then under the action of the controller (11.13), chaos in PMSM can be avoided and the tracking error of the closed-loop controlled system will converge into a sufficient small neighborhood of the origin and all the closed-loop signals are bounded. Moreover, the control properties can avoid the influence of undeterministic parameters.*

Proof To address the stability analysis of the resulting closed-loop system, substitute (11.18) into (11.17) to obtain that

$$\dot{V} \leq -\sum_{i=1}^{3} k_i z_i^2 + \frac{1}{2} l_3^2 + \frac{1}{2} \varepsilon_3^2 - \frac{m_1}{r_1} \tilde{\gamma}\hat{\gamma} - \frac{m_2}{r_2} \tilde{\theta}\hat{\theta}. \tag{11.19}$$

For the term $-\tilde{\gamma}\hat{\gamma}$, one has $-\tilde{\gamma}\hat{\gamma} \leq -\tilde{\gamma}(\tilde{\gamma}+\gamma) \leq -\frac{1}{2}\tilde{\gamma}^2 + \frac{1}{2}\gamma^2$. Similarly, $-\tilde{\theta}\hat{\theta} \leq -\frac{1}{2}\tilde{\theta}^2 + \frac{1}{2}\theta^2$ holds. Consequently, by using these inequalities (11.19) can be rewritten in the following form.

$$\begin{aligned} \dot{V} &\leq -\sum_{i=1}^{3} k_i z_i^2 - \frac{m_1}{2r_1}\tilde{\gamma}^2 - \frac{m_2}{2r_2}\tilde{\theta}^2 + \frac{1}{2} l_3^2 + \frac{1}{2}\varepsilon_3^2 + \frac{m_1}{2r_1}\gamma^2 + \frac{m_2}{2r_2}\theta^2 \\ &\leq -a_0 V + b_0, \end{aligned} \tag{11.20}$$

where $a_0 = \min\left\{2k_1, 2k_2, 2k_3, m_1, m_2\right\}$ and $b_0 = \frac{1}{2}l_3^2 + \frac{1}{2}\varepsilon_3^2 + \frac{m_1}{2r_1}\gamma^2 + \frac{m_2}{2r_2}\theta^2$. Furthermore, (11.20) implies that

$$V(t) \leq (V(t_0) - \frac{b_0}{a_0})e^{-a_0(t-t_0)} + \frac{b_0}{a_0} \leq V(t_0) + \frac{b_0}{a_0}, \forall t \geqslant t_0. \tag{11.21}$$

As a result, all $z_i (i = 1, 2, 3)$, $\tilde{\gamma}$ and $\tilde{\theta}$ belong to the compact set

$$\Omega = \left\{(z_i, \tilde{\gamma}, \tilde{\theta}) | V \leq V(t_0) + \frac{b_0}{a_0}, \forall t \geqslant t_0\right\}.$$

Namely, all the signals in the closed-loop system are bounded. Especially, from (11.21) we have

$$\lim_{t\to\infty} z_1^2 \le \frac{2b_0}{a_0}. \tag{11.22}$$

From the definitions of a_0 and b_0, it is clear that to get a small tracking error by taking r_i sufficiently large and l_i and ε_i small enough after giving the parameters k_i and m_i.

■

11.4 Simulation Results

In this section, we will give compare the proposed approach and the classical backstepping technique. To this end, the classical backstepping is first used to control design for the system (11.2), and the simulation is carried out by both of the proposed method and the classical one.

11.4.1 Classical Backstepping Design

This subsection devotes to design controllers by classical backstepping approach. The control of PMSM based on the conventional backstepping technique is reviewed here.

Step 1: For the reference signal x_d, define the tracking error variable as $z_1 = x_1 - x_d$. From the first differential equation of (11.2), the error dynamic system is given by $\dot{z}_1 = \sigma(x_2 - x_1) - \dot{x}_d$.

Choose Lyapunov function candidate as $V_1 = \frac{1}{2}z_1^2$, then the time derivative of V_1 is computed by

$$\dot{V}_1 = z_1\dot{z}_1 = z_1\sigma(x_2 - x_1 - \frac{\dot{x}_d}{\sigma}). \tag{11.23}$$

Construct the virtual control law α_1 as

$$\alpha_1 = -\bar{k}_1 z_1 + x_1 + \frac{\dot{x}_d}{\sigma}, \tag{11.24}$$

with $\bar{k}_1$ is a positive constant. By using (11.24), (11.23) can be rewritten of the following form.

$$\dot{V}_1 = -\bar{k}_1\sigma z_1^2 + \sigma z_1 z_2 = -k_1 z_1^2 + \sigma z_1 z_2,$$

with $k_1 = \bar{k}_1\sigma > 0$ being a design parameter and $z_2 = x_2 - \alpha_1$.

Step 2: Differentiating z_2 gives

$$\dot{z}_2 = \dot{x}_2 - \dot{\alpha}_1 = -x_2 - x_1 x_3 + \gamma x_1 - \dot{\alpha}_1. \tag{11.25}$$

Now, choose the Lyapunov function candidate as $V_2 = V_1 + \frac{1}{2}z_2^2$. Obviously, the time derivative of V_2 is given by

$$\dot{V}_2 = \dot{V}_1 + z_2\dot{z}_2 = -k_1 z_1^2 + z_2(\sigma z_1 - x_2 - x_1 x_3 + \gamma x_1 - \dot{\alpha}_1). \tag{11.26}$$

The virtual control α_2 is constructed as

$$\alpha_2 = -\frac{1}{x_1}(-k_2 z_2 - \sigma z_1 + x_2 - \gamma x_1 + \dot{\alpha}_1), \tag{11.27}$$

where $k_2 > 0$ is a positive design parameter and $\dot{\alpha}_1 = \dot{x}_1 - \dot{x}_d = \sigma(x_2 - x_1) - \dot{x}_d$. Adding and subtracting α_2 in the bracket in (11.26) shows that

$$\dot{V}_2 = -k_1 z_1^2 - k_2 z_2^2 - x_1 z_2 z_3, \tag{11.28}$$

with $z_3 = x_3 - \alpha_2$.

Step 3: Differentiating z_3 results in the following differential equation

$$\dot{z}_3 = \dot{x}_3 - \dot{\alpha}_2 = -x_3 + x_1 x_2 + u_d - \dot{\alpha}_2.$$

Choose the Lyapunov function candidate as $V_3 = V_2 + \frac{1}{2}z_3^2$. Furthermore, differentiating V_3 yields

$$\begin{aligned}\dot{V}_3 &= \dot{V}_2 + z_3\dot{z}_3 = \dot{V}_2 + z_3(-x_3 + x_1 x_2 + u_d - \dot{\alpha}_2)\\ &= -k_1 z_1^2 - k_2 z_2^2 + z_3(-x_3 + x_1 x_2 - x_1 z_2 - \dot{\alpha}_2 + u_d),\end{aligned} \tag{11.29}$$

where

$$\begin{aligned}\dot{\alpha}_2 &= \sum_{i=1}^{2}\frac{\partial\alpha_2}{\partial x_i}\dot{x}_i + \sum_{i=0}^{2}\frac{\partial\alpha_2}{\partial x_d^{(i)}}x_d^{(i+1)}\\ &= \frac{\partial\alpha_2}{\partial x_1}\sigma(x_2 - x_1) + \frac{\partial\alpha_2}{\partial x_2}(-x_2 - x_1 x_3 + \gamma x_1) + \sum_{i=0}^{2}\frac{\partial\alpha_2}{\partial x_d^{(i)}}x_d^{(i+1)}.\end{aligned}$$

Then the control input u_d is designed as

$$\begin{aligned}u_d &= -k_3 z_3 + x_3 - x_1 x_2 + x_1 z_2 + \dot{\alpha}_2\\ &= -k_3 z_3 + x_3 - x_1 x_2 + x_1 z_2 + \frac{\partial\alpha_2}{\partial x_1}\sigma(x_2 - x_1)\\ &\quad + \frac{\partial\alpha_2}{\partial x_2}(-x_2 - x_1 x_3 + \gamma x_1) + \sum_{i=0}^{2}\frac{\partial\alpha_2}{\partial x_d^{(i)}}x_d^{(i+1)},\end{aligned} \tag{11.30}$$

where $k_3 > 0$.

11.4.2 Simulation

In order to illustrate the effectiveness of the proposed results, the simulation will be conducted to control chaos in PMSM drive system under the initial condition of $x_1 = x_2 = x_3 = 0.5$. Firstly, we tested the chaotic PMSM drive system with $u_d = 0$, shown in Figs. 11.2, 11.3 and 11.4. Secondly, the proposed adaptive fuzzy approach in this paper is used to control the chaotic PMSM system, shown in Figs. 11.5, 11.6 and 11.7. Thirdly, the curve of the virtual controllers α_1, α_2 and the final controller u_d are shown in Figs. 11.8 and 11.9. Moreover, the robustness of the proposed controller against uncertainty in system parameters is analyzed by simulation. Figure 11.10 shows the curves of undeterministic parameter estimate error $(\hat{\gamma} - \gamma)$ according to its corresponding adaptation law. The control parameters are chosen as follows:

$$k_1 = 2, \quad k_2 = 20, \quad k_3 = 15, \quad r_1 = r_2 = 15, \quad m_1 = m_2 = 0.005, \quad l_3 = 0.2.$$

And the fuzzy membership functions are:

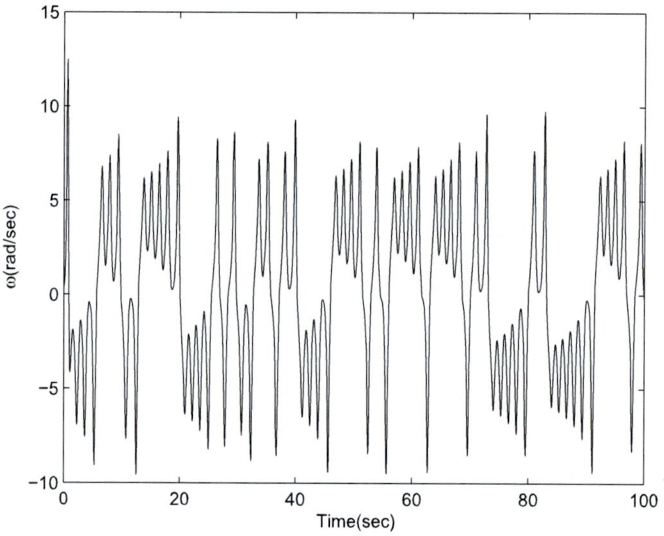

Fig. 11.2 Curve of the ω for chaotic PMSM drive system without u_d

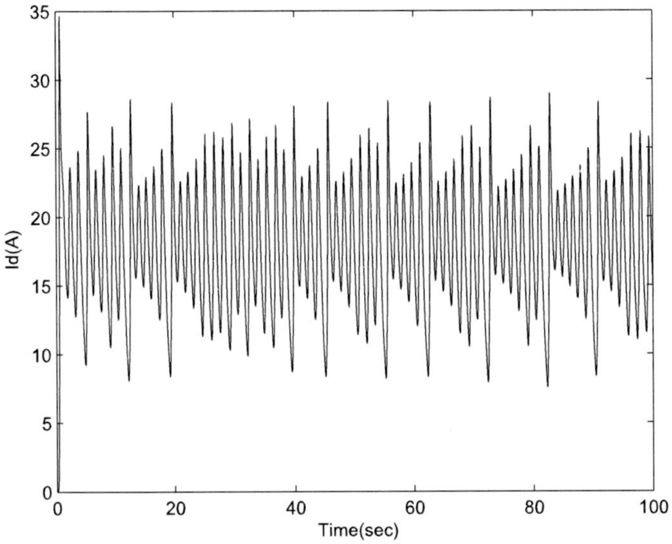

Fig. 11.3 Curve of the i_d for chaotic PMSM drive system without u_d

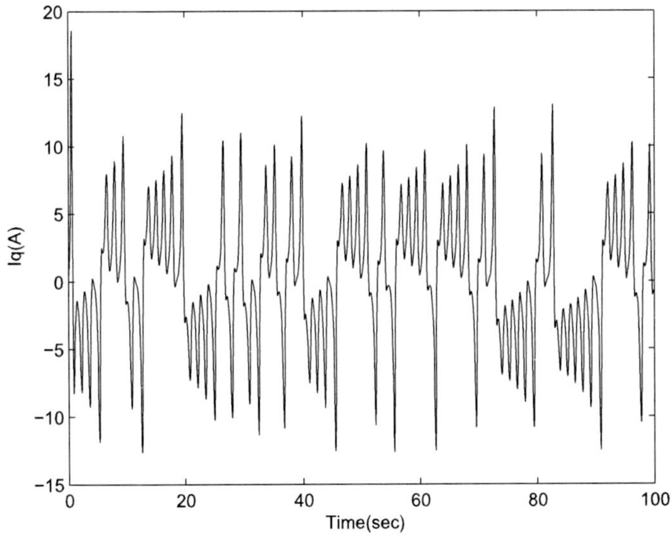

Fig. 11.4 Curve of the i_q for chaotic PMSM drive system without u_d

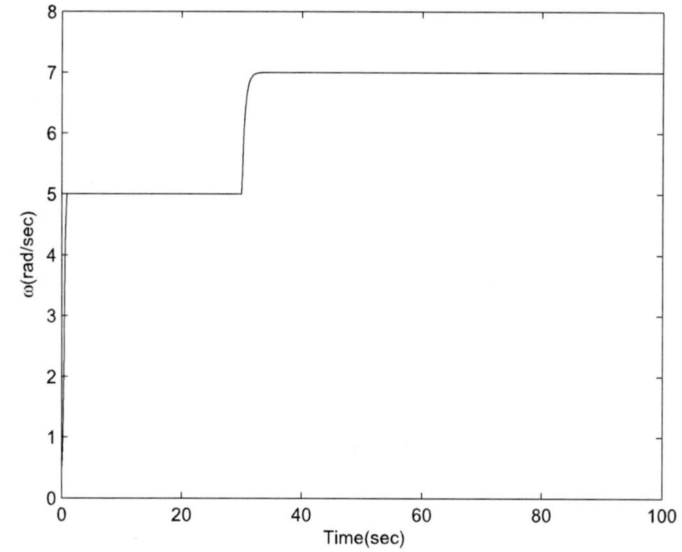

Fig. 11.5 Curve of the ω for chaotic PMSM drive system when utilizing the controller u_d

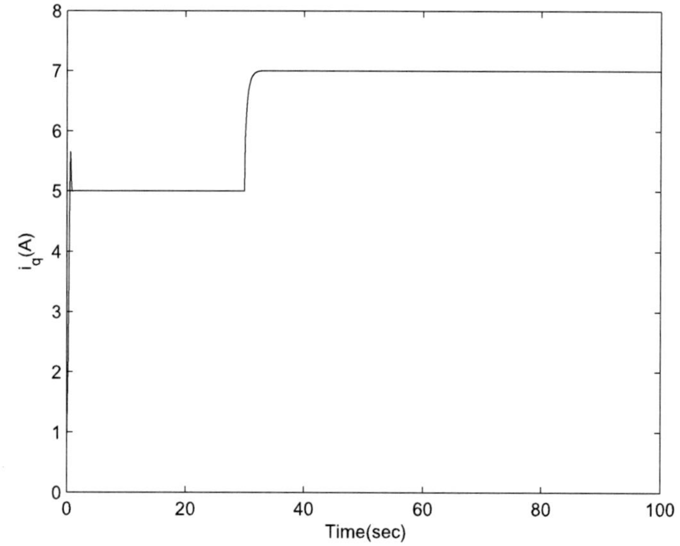

Fig. 11.6 Curve of the i_q for PMSM drive system when utilizing the controller u_d

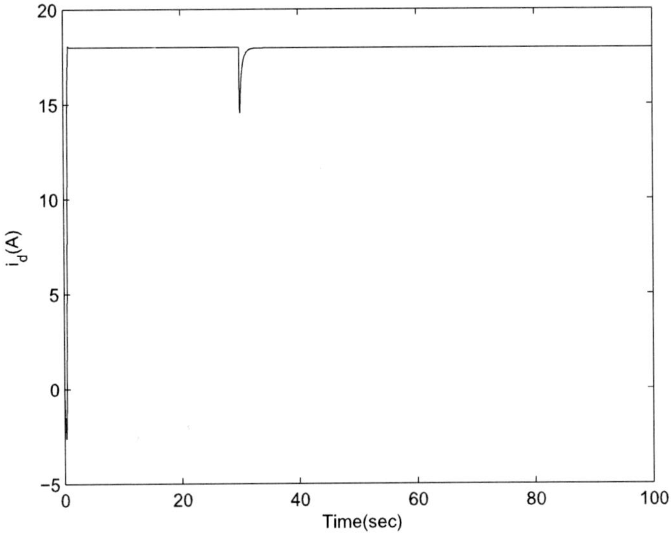

Fig. 11.7 Curve of the i_d for chaotic PMSM drive system when utilizing the controller u_d

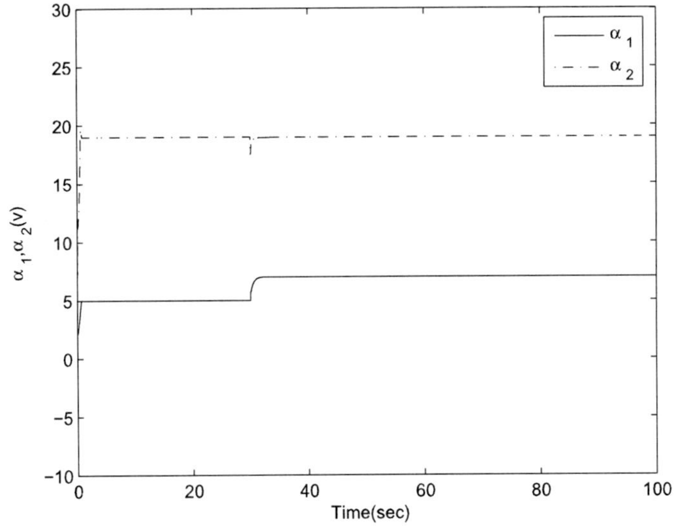

Fig. 11.8 Curves of the virtual controller α_1 and α_2

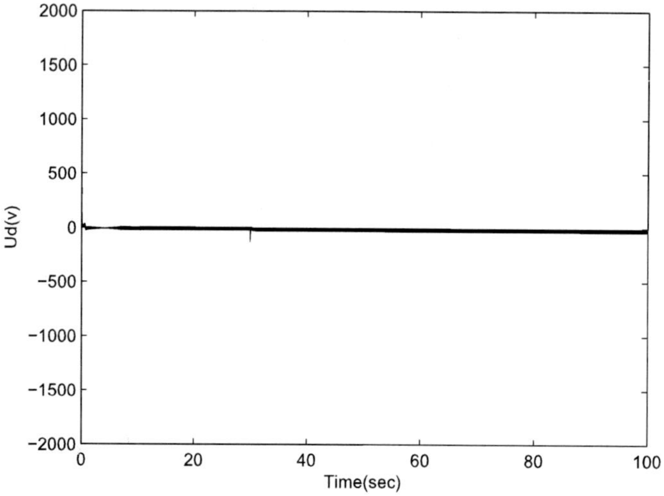

Fig. 11.9 Curve of the controller u_d

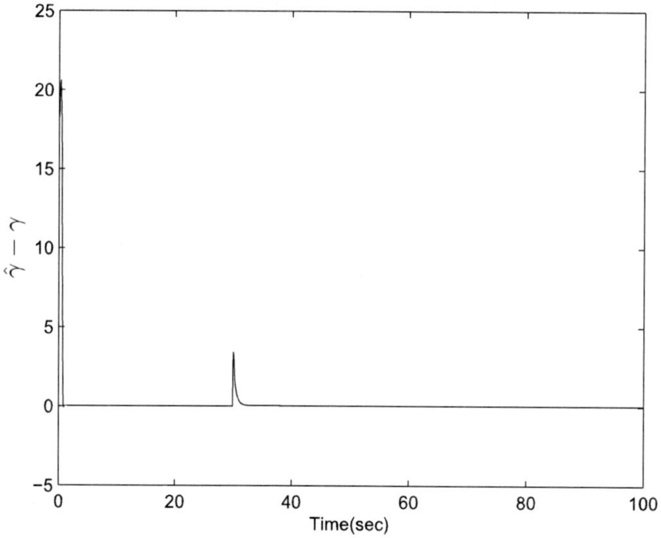

Fig. 11.10 Curve of the parameter estimate error $(\hat{\gamma} - \gamma)$

$$\mu_{F_i^1} = \exp\left[\frac{-(x+7)^2}{2}\right], \quad \mu_{F_i^2} = \exp\left[\frac{-(x+6)^2}{2}\right],$$
$$\mu_{F_i^3} = \exp\left[\frac{-(x+5)^2}{2}\right], \quad \mu_{F_i^4} = \exp\left[\frac{-(x+4)^2}{2}\right],$$
$$\mu_{F_i^5} = \exp\left[\frac{-(x+3)^2}{2}\right], \quad \mu_{F_i^6} = \exp\left[\frac{-(x+2)^2}{2}\right],$$
$$\mu_{F_i^7} = \exp\left[\frac{-(x+1)^2}{2}\right], \quad \mu_{F_i^8} = \exp\left[\frac{-(x-0)^2}{2}\right],$$
$$\mu_{F_i^9} = \exp\left[\frac{-(x-1)^2}{2}\right], \quad \mu_{F_i^{10}} = \exp\left[\frac{-(x-2)^2}{2}\right],$$
$$\mu_{F_i^{11}} = \exp\left[\frac{-(x-3)^2}{2}\right], \quad \mu_{F_i^{12}} = \exp\left[\frac{-(x-4)^2}{2}\right],$$
$$\mu_{F_i^{13}} = \exp\left[\frac{-(x-5)^2}{2}\right], \quad \mu_{F_i^{14}} = \exp\left[\frac{-(x-6)^2}{2}\right],$$
$$\mu_{F_i^{15}} = \exp\left[\frac{-(x-7)^2}{2}\right].$$

Give the reference signal

$$x_d = \begin{cases} 5, & 0 \le t \le 20, \\ 7, & t \ge 20, \end{cases}$$

and the simulation is carried out for the PMSM drive system. It is seen clearly that the proposed controller can suppress the chaos in the PMSM drive system and good tracking performance has been achieved successfully.

Remark 11.5 Notice $5 \le |x_d| \le 7$ and the system's output, i.e., x_1 will follow x_d under the action of the control input. This means that we can choose the interval [−7, 7] for the state variable x_1. For simplicity in simulation, this interval is still used for the other state variables. Furthermore, according to Lemma 11.1 when we chose the membership function to be a kind of Gaussian functions and cover this interval, the corresponding fuzzy logic systems can approximate the nonlinear functions defined on this interval. ♦

11.5 Conclusion

Based on the backstepping technique, an adaptive fuzzy tracking control method is proposed to suppress chaos in the permanent magnet synchronous motor drive systems based on the backstepping technique. The proposed controller that solves the traditional backstepping control's main problems guarantees that the tracking error converges to a small neighborhood of the origin and all the closed-loop signals are bounded. The simulation results are supplied to demonstrate the effectiveness and robustness against the parameter uncertainties in a chaotic drive system.

References

1. Ott, E., Grebogi, C., Yorke, J.A.: Controlling chaos. Phys. Rev. Lett. **64**, 1196–1199 (1990)
2. Li, Z., Chen, G., Shi, S., Han. C.: Robust adaptive tracking control for a class of uncertain chaotic systems. Phys. Lett. A **310**, 40–43 (2003)
3. Arecchi, F.T., Boccaletti, S.: Adaptive strategies for recognition, noise filtering, control, synchronization and targeting of chaos. Chaos **7**, 621–634 (1997)
4. Boccaletti, S., Grebogi, C., Lai, Y.C., Mancini, H., Maza, D.: The control of chaos: theory and applications. Phys. Rep. **329**, 103–197 (2000)
5. Kurths, J., Boccaletti, S., Grebogi, C., Lai, Y.C.: Introduction: control and synchronization in chaotic dynamical systems. Chaos **13**, 126–127 (2003)
6. Kuroe, Y., Hayash, S.: Analysis of bifurcation in power electronic induction motor drive system. IEEE Power Elect. Speci. Confe. Rec. 923–930 (1989)
7. Chen, J.H., Chau, K.T., Siu, S.M., Chan, C.C.: Experimental stabilization of chaos in a voltage-mode dc drive system. IEEE Trans. Circuits Syst. I **47**(7), 1093–1095 (2000)
8. Wang, Z., Chau, K.T.: Anti-control of chaos of a permanent magnet dc motor system for vibratory compactors. Chaos Solitons Fractals **36**, 694–708 (2008)
9. Ge, Z.M., Chang, C.M., Chen, Y.S.: Anti-control of chaos of single time scale brushless dc motors and chaos synchronization of different order systems. Chaos Solitons Fractals **27**, 1298–1315 (2006)
10. Robert, B., Alin, F., Goeldel, C.: Aperiodic and chaotic dynamics in hybrid step motor-new experimental results. In: Proceedings of the IEEE International Symposium on Industrial Electronics, Pusan, Korea, vol. 3, pp. 2136–2141 (2001)
11. Gao, Y., Chau, K.T., Ye, S.: A novel chaotic-speed single-phase induction motor drive for cooling fans. In: The 40th IAS Annual Meeting on Industry Applications Conference, vol. 2, pp. 1337–1341 (2005)
12. Harb, A.M.: Nonlinear chaos control in a permanent magnet reluctance machine. Chaos Solitons Fractals **19**, 1217–1224 (2004)
13. Gao, Y., Chau, K.T.: Hopf bifurcation and chaos in synchronous reluctance motor drives. IEEE Trans. Energ. Conve. **19**(2), 296–302 (2004)
14. Chen, J.H., Chau, K.T., Jiang, Q., Chan, C.C., Jiang, S.Z.: Modeling and analysis of chaotic behavior in switched reluctance motor drives. In: IEEE 31st Annual Power Electronics Specialists Conference, vol. 3, pp. 1551–1556 (2000)
15. Li, Z., Park, J.B., Joo, Y.H., Zhang, B., Chen, G.: Bifurcations and chaos in a permanent-magnet synchronous motor. IEEE Trans. Circuits Syst. I Fundam. Theory Appl. **49**, 383–387 (2002)
16. Wei, D.Q., Luo, X.S., Wang, B.H., Fang, J.Q.: Robust adaptive dynamic surface control of chaos in permanent magnet synchronous motor. Phys. Lett. A **363**, 71–77 (2007)
17. Zribi, M., Oteafy, A., Smaoui, N.: Controlling chaos in the permanent magnet synchronous motor. Chaos Solitons Fractals **41**(3), 1266–1276 (2009)
18. Wang, L.X., Mendel, J.M.: Fuzzy basis functions, universal approximation, and orthogonal least squares learning. IEEE Trans. Neural Netw. **3**(5), 807–814 (1992)
19. Stotsky, A., Hedrick, J.K., Yip, P.P.: The use of sliding modes to simplify the backstepping control method. Proc. Am. Control Conf. **3**, 1703–1708 (1997)

Chapter 12
Fuzzy-Approximation-Based Adaptive Control of the Chaotic PMSM

The problem of position tracking control for the chaotic permanent magnet synchronous motor drive system with parameter uncertainty is studied in this chapter. The fuzzy logic systems are used to approximate the nonlinearities and the adaptive backstepping technique is used to construct the controllers. The tracking error can converge to a small neighborhood of the origin through the proposed adaptive fuzzy controllers. And the structure of designed fuzzy controllers is extremely simple compared with the conventional backstepping. Simulation results show that the proposed control scheme can suppress the chaos of PMSM and retain the perfect tracking performance under unknown parameters.

12.1 Introduction

Permanent magnet synchronous motor (PMSM) is widely concerned in industrial applications because of its high speed, high efficiency, high power density, and large torque to inertia ratio. The secure and stable operation of the PMSM, which is a vital requirement of industrial automation manufacturing, has accepted considerable attention due to its dynamic model is nonlinear, multivariable, and even experiencing Hopf bifurcation, limit cycles, and chaotic attractors with systemic parameters falling into a certain area. Chaotic behavior [1–3] of the PMSM is not desirable, it will destroy the stability of the motor, and even cause system collapse. Since Kuroe and Hayashi [4] proposed the occurrence of chaos in motor drive system in the late 1980s, many researchers have begun to pay attention to the discovery and control of chaos in motor operation and the references are as follows [5–11].

An adaptive fuzzy approximation method is proposed to suppress chaos in PMSM drive system by backstepping technology [12–14] in this chapter. In controller design, fuzzy logic systems are used to approximate the nonlinearity of the chaotic PMSM

J. Yu et al., *Intelligent Backstepping Control for the Alternating-Current Drive Systems*, Studies in Systems, Decision and Control 349,
https://doi.org/10.1007/978-3-030-67723-7_12

drive system, and the "term explosion" problem is overcome. In order to verify the superiority of the proposed control method, the two methodologies are compared. In addition, the proposed controllers ensure that the tracking error converges to a small neighborhood of the origin, and all the closed-loop signals are bounded.

12.2 Mathematical Model of Chaotic PMSM Drive System and Preliminaries

The dimensionless mathematical model of PMSM with the smooth air gap can be described as follows [5]:

$$\begin{aligned}\frac{d\theta}{dt} &= \omega,\\ \frac{d\omega}{dt} &= \sigma(i_q - \omega) - \tilde{T}_L,\\ \frac{di_q}{dt} &= -i_q - i_d\omega + \gamma\omega + \tilde{u}_q,\\ \frac{di_d}{dt} &= -i_d + i_q\omega + \tilde{u}_d,\end{aligned} \tag{12.1}$$

where θ, ω, i_d and i_q are state variables, which denote the rotor position, angle speed and the $d-q$ axis currents. σ and γ are system operating parameters, which are positive. $\tilde{T}_L$, $\tilde{u}_d$ and $\tilde{u}_q$ stand for the $d-q$ axis voltages and load torque.

In the system (12.1), the external inputs are set to zero, namely, $\tilde{T}_L = \tilde{u}_d = \tilde{u}_q = 0$. Then, the system (12.1) becomes an unforced system:

$$\begin{aligned}\frac{d\theta}{dt} &= \omega,\\ \frac{d\omega}{dt} &= \sigma(i_q - \omega),\\ \frac{di_q}{dt} &= -i_q - i_d\omega + \gamma\omega,\\ \frac{di_d}{dt} &= -i_d + i_q\omega.\end{aligned}$$

It is found that when the operating parameters σ and γ fall into a certain area [5], the PMSM is experiencing chaotic behavior. For example, the PMSM begins to display chaos with $\sigma = 5.46$ and $\gamma = 14.93$. The typical chaotic attractor is shown in Fig. 12.1. The chaotic oscillations can destroy the stabilization of the PMSM drive system.

Remark 12.1 In this chapter, we assume the parameter σ is unknown, but its lower bound is 1, namely, $\sigma \geq 1$. Therefore, the proposed control method is suitable for any σ, which is larger than 1. ♦

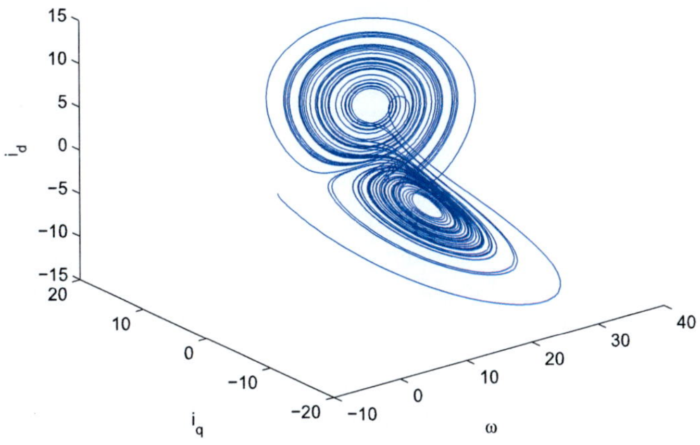

Fig. 12.1 Curves of the typical chaotic attractor in PMSM with system parameters

In order to eliminate chaos in the PMSM drive system, we use u_d and u_q as the manipulated variable which is desirable for the real application. An adaptive fuzzy control approach is proposed to suppress chaos via backstepping. For the sake of simplicity, we introduce the following notations : $x_1 = \theta$, $x_2 = \omega$, $x_3 = i_q$, $x_4 = i_d$. By using these notations, the dynamic model of the PMSM driver system can be described by the following differential equations:

$$\begin{aligned}
\dot{x}_1 &= x_2, \\
\dot{x}_2 &= \sigma(x_3 - x_2), \\
\dot{x}_3 &= -x_3 - x_2 x_4 + \gamma x_2 + u_q, \\
\dot{x}_4 &= -x_4 + x_2 x_3 + u_d.
\end{aligned} \tag{12.2}$$

The control objective is to design an adaptive fuzzy controller such that the state variable x_1 follows the given reference signal x_d and all the closed-loop signals are bounded.

Lemma 12.2 [15] *Let $f(x)$ be a continuous function defined on a compact set Ω. Then for any scalar $\varepsilon > 0$, there exists a fuzzy logic system $W^T S(x)$ such that*

$$\sup_{x \in \Omega} \left| f(x) - W^T S(x) \right| \le \varepsilon,$$

where $W = [W_1, \ldots, W_N]^T$ is the ideal constant weight vector, and $S(x) = [p_1(x), p_2(x), \ldots, p_N(x)]^T / \sum_{i=1}^{N} p_i(x)$ is the basis function vector, with $N > 1$ being the number of the fuzzy rules and p_i are chosen as Gaussian functions, i.e., for

$i = 1, 2, \ldots, N$, $p_i(x) = \exp[\frac{-(x-\mu_i)^T(x-\mu_i)}{\eta_i^2}]$ *where* $\mu_i = [\mu_{i1}, \mu_{i2}, \ldots, \mu_{in}]^T$ *is the center vector, and* η_i *is the width of the Gaussian function.*

12.3 Adaptive Fuzzy Controller with the Backstepping Technique

In this section, we propose an adaptive fuzzy control method to control chaos in the PMSM drive system. The backstepping design process consists of four steps. At each design step, a virtual control function α_i $(i = 1, 2)$ will be constructed by using an appropriate Lyapunov function. At the last step, the real controller is constructed to control the system.

Step 1: For the reference signal x_d, define the tracking error variable as $z_1 = x_1 - x_d$. From the first differential equation of (12.2), the error dynamic system is given by $\dot{z}_1 = x_2 - \dot{x}_d$.

The Lyapunov function is chosen as $V_1 = \frac{1}{2}z_1^2$, then the time derivative of V_1 is computed by

$$\dot{V}_1 = z_1\dot{z}_1 = z_1(x_2 - \dot{x}_d). \tag{12.3}$$

Construct the virtual control law α_1 as

$$\alpha_1(x_1, x_d, \dot{x}_d) = -k_1z_1 + \dot{x}_d, \tag{12.4}$$

with $k_1 > 0$ being a design parameter and $z_2 = x_2 - \alpha_1$. By using (12.4), (12.3) can be rewritten of the following form.

$$\dot{V}_1 = -k_1z_1^2 + z_1z_2.$$

Step 2: Differentiating z_2 gives

$$\dot{z}_2 = \dot{x}_2 - \dot{\alpha}_1 = \sigma(x_3 - x_2) - \dot{\alpha}_1. \tag{12.5}$$

Now, the Lyapunov function is chosen as $V_2 = V_1 + \frac{1}{2}z_2^2$. Obviously, V_2 's derivative can be written as

$$\begin{aligned}\dot{V}_2 = \dot{V}_1 + z_2\dot{z}_2 &= -k_1z_1^2 + z_2(z_1 + \sigma(x_3 - x_2) - \dot{\alpha}_1) \\ &= -k_1z_1^2 + (\sigma z_2x_3 + z_2f_2),\end{aligned} \tag{12.6}$$

where $f_2(Z_2) = -\sigma x_2 + z_1 - \dot{\alpha}_1$, $Z_2 = [x_1, x_2, x_d, \dot{x}_d, \ddot{x}_d]^T$. Notice that f_2 contains the derivative of α_1, which is a nonlinear term and the parameter σ is unknown, this will make the classical adaptive backstepping design become troublesome. In order to solve this problem, we will employ the fuzzy logic system to approximate the nonlinear function f_2. According to Lemma 12.2, for any given $\varepsilon_2 > 0$, there exists a fuzzy logic system $W_2^T S_2(Z_2)$ such that

$$f_2(Z_2) = W_2^T S_2(Z_2) + \delta_2(Z_2), \tag{12.7}$$

where $\delta_2(Z_2)$ is the approximation error and satisfies $|\delta_2| \leq \varepsilon_2$. Consequently, a straightforward calculation produces the following inequality.

$$\begin{aligned} z_2 f_2 = z_2\left(W_2^T S_2 + \delta_2\right) &\leq \frac{1}{2l_2^2} z_2^2 \left\|W_2\right\|^2 S_2^T S_2 + \frac{1}{2} l_2^2 + \frac{1}{2} z_2^2 + \frac{1}{2}\varepsilon_2^2 \\ &\leq \frac{\|W_2\|^2}{2l_2^2} z_2^2 + \frac{1}{2} l_2^2 + \frac{1}{2} z_2^2 + \frac{1}{2}\varepsilon_2^2, \end{aligned} \tag{12.8}$$

where the inequality $S_2^T S_2 \leq 1$ is used. Thus, it follows immediately from substituting (12.8) into (12.6) that

$$\dot{V}_2 \leq -k_1 z_1^2 + \frac{1}{2l_2^2} z_2^2 \|W_2\|^2 + \frac{1}{2} l_2^2 + \frac{1}{2} z_2^2 + \frac{1}{2}\varepsilon_2^2 + \sigma z_2 x_3. \tag{12.9}$$

The virtual control α_2 is constructed as

$$\alpha_2(x_1, x_2, x_d, \dot{x}_d, \ddot{x}_d) = -\frac{1}{\sigma}[(k_2 + \frac{1}{2}) z_2 + \frac{1}{2l_2^2} z_2 \hat{\phi}], \tag{12.10}$$

where $\hat{\phi}$ is the estimation of the unknown constant ϕ which will be specified later. Adding and subtracting α_2 in (12.9) shows that

$$\begin{aligned} \dot{V}_2 \leq& -k_1 z_1^2 + \frac{1}{2l_2^2} z_2^2 \|W_2\|^2 + \frac{1}{2} l_2^2 + \frac{1}{2} z_2^2 \\ &+ \frac{1}{2}\varepsilon_2^2 + \sigma z_2(-\frac{1}{\sigma}[(k_2 + \frac{1}{2}) z_2 + \frac{1}{2l_2^2} z_2 \hat{\phi}] + z_3) \\ \leq& -k_1 z_1^2 - k_2 z_2^2 + \frac{1}{2l_2^2} z_2^2 (\|W_2\|^2 - \hat{\phi}) + \frac{1}{2} l_2^2 + \frac{1}{2}\varepsilon_2^2 + \sigma z_2 z_3, \end{aligned} \tag{12.11}$$

with $k_2 > 0$ being a design parameter and $z_3 = x_3 - \alpha_2$.

Step 3: Differentiating z_3 results in the following differential equation

$$\dot{z}_3 = \dot{x}_3 - \dot{\alpha}_2 = -x_3 - x_2 x_4 + \gamma x_2 + u_q - \dot{\alpha}_2.$$

Choose the Lyapunov function candidate as $V_3 = V_2 + \frac{1}{2} z_3^2$. Furthermore, differentiating V_3 yields

$$\begin{aligned}\dot{V}_3 &= \dot{V}_2 + z_3\dot{z}_3 = \dot{V}_2 + z_3(-x_3 - x_2x_4 + \gamma x_2 + u_q - \dot{\alpha}_2)\\ &\le -k_1z_1^2 - k_2z_2^2 + \frac{1}{2l_2^2}z_2^2(\|W_2\|^2 - \hat{\phi})\\ &\quad + \frac{1}{2}l_2^2 + \frac{1}{2}\varepsilon_2^2 + z_3(f_3 + u_q),\end{aligned} \tag{12.12}$$

where

$$\begin{aligned}\dot{\alpha}_2 &= \sum_{i=1}^{2}\frac{\partial\alpha_2}{\partial x_i}\dot{x}_i + \sum_{i=0}^{2}\frac{\partial\alpha_2}{\partial x_d^{(i)}}x_d^{(i+1)}\\ &= \frac{\partial\alpha_2}{\partial x_1}x_2 + \frac{\partial\alpha_2}{\partial x_2}\sigma(x_3 - x_2) + \sum_{i=0}^{2}\frac{\partial\alpha_2}{\partial x_d^{(i)}}x_d^{(i+1)},\\ f_3(Z_3) &= -x_3 - x_2x_4 + \gamma x_2 + \sigma z_2 - \dot{\alpha}_2,\\ Z_3 &= [x_1, x_2, x_3, x_4, x_d, \dot{x}_d, \ddot{x}_d]^T.\end{aligned} \tag{12.13}$$

Due to the limitation of working conditions, the parameter γ is usually unknown in the real model of PMSM. So it cannot be used to construct the control signal. Since the derivative of α_2 and $-x_2x_4$ are nonlinear terms in f_3, this will make the classical adaptive backstepping design become very complex and troubled, and the designed control law u_q will have a complex structure. Similarly, the fuzzy logic system $W_3^T S_3(Z_3)$ is utilized to approximate the nonlinear function f_3. As shown later, the design procedure of u_q becomes simple and u_q is of a simple structure. For given $\varepsilon_3 > 0$, the following equation holds

$$z_3f_3 = z_3\left(W_3^T S_3 + \delta_3\right) \le \frac{1}{2l_3^2}z_3^2\,\|W_3\|^2 + \frac{1}{2}l_3^2 + \frac{1}{2}z_3^2 + \frac{1}{2}\varepsilon_3^2. \tag{12.14}$$

Thus, it follows immediately from substituting (12.14) into (12.12) that

$$\begin{aligned}\dot{V}_3 &\le -k_1z_1^2 - k_2z_2^2 + \frac{1}{2l_2^2}z_2^2(\|W_2\|^2 - \hat{\phi}) + \frac{1}{2}z_3^2\\ &\quad + \sum_{i=2}^{3}\frac{1}{2}(l_i^2 + \varepsilon_i^2) + \frac{1}{2l_3^2}z_3^2\,\|W_3\|^2 + z_3u_q.\end{aligned} \tag{12.15}$$

Now, the control law u_q is designed as

$$u_q = -(k_3 + \frac{1}{2})z_3 - \frac{1}{2l_3^2}z_3\hat{\phi}. \tag{12.16}$$

Furthermore, by using the equality (12.16), it can be verified easily that

$$\dot{V}_3 \le -\sum_{i=1}^{3} k_i z_i^2 + \sum_{i=2}^{3} \frac{1}{2}(l_i^2 + \varepsilon_i^2) + \sum_{i=2}^{3} \frac{1}{2l_i^2} z_i^2 \left(\|W_i\|^2 - \hat{\phi} \right).$$

Step 4: At this step, the control law u_d will be constructed. To this end, define $z_4 = x_4$ and choose the following Lyapunov function candidate as $V_4 = V_3 + \frac{1}{2}z_4^2$. Then the derivative of V_4 is given by

$$\begin{aligned} \dot{V}_4 &= \dot{V}_3 + z_4 \dot{z}_4 \\ &\le -\sum_{i=1}^{3} k_i z_i^2 + \sum_{i=2}^{3} \frac{1}{2l_i^2} z_i^2 \left(\|W_i\|^2 - \hat{\phi} \right) \\ &\quad + \sum_{i=2}^{3} \frac{1}{2}(l_i^2 + \varepsilon_i^2) + z_4(f_4 + u_d), \end{aligned} \tag{12.17}$$

where $f_4(Z_4) = -x_4 + x_2 x_3$. By Lemma 12.2 the fuzzy logic system $W_4^T S_4(Z_4)$ is utilized to approximate the nonlinear function f_4 such that for given $\varepsilon_4 > 0$,

$$z_4 f_4 \le \frac{1}{2l_4^2} z_4^2 \|W_4\|^2 + \frac{1}{2} l_4^2 + \frac{1}{2} z_4^2 + \frac{1}{2} \varepsilon_4^2. \tag{12.18}$$

Combining (12.18) and (12.17) gives

$$\begin{aligned} \dot{V}_4 &= \dot{V}_3 + z_4 \dot{z}_4 \\ &\le -\sum_{i=1}^{3} k_i z_i^2 + \sum_{i=2}^{4} \frac{1}{2}(l_i^2 + \varepsilon_i^2) + \frac{1}{2l_4^2} z_4^2 \|W_4\|^2 \\ &\quad + \sum_{i=2}^{3} \frac{1}{2l_i^2} z_i^2 \left(\|W_i\|^2 - \hat{\phi} \right) + \frac{1}{2} z_4^2 + z_4 u_d. \end{aligned} \tag{12.19}$$

Now choose u_d as

$$u_d = -(k_4 + \frac{1}{2}) z_4 - \frac{1}{2l_4^2} z_4 \hat{\phi}, \tag{12.20}$$

and define $\phi = \max\{\|W_2\|^2, \|W_3\|^2, \|W_4\|^2\}$. Then, combining (12.19) with (12.20) results in

$$\dot{V}_4 \le -\sum_{i=1}^{4} k_i z_i^2 + \sum_{i=2}^{4} \frac{1}{2}(l_i^2 + \varepsilon_i^2) + \sum_{i=2}^{4} \frac{1}{2l_i^2} z_i^2 \left(\phi - \hat{\phi} \right). \tag{12.21}$$

Introduce variables $\tilde{\phi}$ as

$$\tilde{\phi} = \hat{\phi} - \phi, \tag{12.22}$$

and choose the Lyapunov function candidate as:

$$V = V_4 + \frac{1}{2r_1}\tilde{\phi}^2, \tag{12.23}$$

where r_1 is a positive constant. By differentiating V and taking (12.21)–(12.23) into account, one has

$$\begin{aligned}\dot{V} &\leq -\sum_{i=1}^{4} k_i z_i^2 + \sum_{i=2}^{4} \frac{1}{2}(l_i^2 + \varepsilon_i^2) + \sum_{i=2}^{4} \frac{1}{2l_i^2} z_i^2 \left(\phi - \hat{\phi}\right) + \frac{1}{r_1}\tilde{\phi}\dot{\hat{\phi}} \\ &= -\sum_{i=1}^{4} k_i z_i^2 + \sum_{i=2}^{4} \frac{1}{2}(l_i^2 + \varepsilon_i^2) + \frac{1}{r_1}\tilde{\phi}\left[-\sum_{i=2}^{4} \frac{r_1}{2l_i^2} z_i^2 + \dot{\hat{\phi}}\right]. \end{aligned} \tag{12.24}$$

According to (12.24), the corresponding adaptive laws are chosen as follows:

$$\dot{\hat{\phi}} = \sum_{i=2}^{4} \frac{r_1}{2l_i^2} z_i^2 - m_1\hat{\phi}, \tag{12.25}$$

where m_1 and l_i, for $i = 2, 3, 4$ are positive constants.

Remark 12.3 Apparently, the proposed fuzzy controllers have simple form. This means that such controllers are easily implemented in practice. To give a comparison with the conventional backstepping controllers, we develop the controllers in equations (12.16) and (12.20) via conventional backstepping. It can be seen clearly that the expression of these controllers (12.36) and (12.39) are much more complicated than these adaptive fuzzy controllers (12.16) and (12.20). The number of terms in the expression of (12.36) and (12.39) are much larger. This drawback is called the "explosion of terms" above [16]. ♦

Remark 12.4 In the realistic model of PMSM, the system parameters σ and γ may be unknown, so, they cannot be used to construct the control signal unless we specify their corresponding adaptation laws. In this chapter, fuzzy logic systems are employed to approximate nonlinearities, so no regression matrices need to be found. Since the unknown σ and γ are the parameters of the nonlinear functions, the undeterministic parameters are taken into account, and we need not specify their corresponding adaptation laws. Thus, the major problems with traditional backstepping are cured. The stability of the system is given by the following theorem. ♦

Theorem 12.5 *Consider the system (12.2) and the reference signal x_d. Then under the action of the controllers (12.16) and (12.20), chaos in PMSM can be avoided and the tracking error of the closed-loop controlled system will converge into a sufficient small neighborhood of the origin and all the closed-loop signals are bounded. Moreover, the control properties can avoid the influence of undeterministic parameters.*

Proof To address the stability analysis of the resulting closed-loop system, substitute (12.25) into (12.24) to obtain that

$$\dot{V} \leq -\sum_{i=1}^{4} k_i z_i^2 + \sum_{i=2}^{4} \frac{1}{2}(l_i^2 + \varepsilon_i^2) - \frac{m_1}{r_1}\tilde{\phi}\hat{\phi}. \tag{12.26}$$

For the term $-\tilde{\phi}\hat{\phi}$, one has $-\tilde{\phi}\hat{\phi} \leq -\tilde{\phi}(\tilde{\phi} + \phi) \leq -\frac{1}{2}\tilde{\phi}^2 + \frac{1}{2}\phi^2$. Consequently, by using these inequalities (12.26) can be rewritten in the following form

$$\begin{aligned} \dot{V} &\leq -\sum_{i=1}^{4} k_i z_i^2 - \frac{m_1}{2r_1}\tilde{\phi}^2 + \sum_{i=2}^{4} \frac{1}{2}(l_i^2 + \varepsilon_i^2) + \frac{m_1}{2r_1}\phi^2 \\ &\leq -a_0 V + b_0, \end{aligned} \tag{12.27}$$

where $a_0 = \min\left\{2k_1, 2k_2, 2k_3, 2k_4, m_1\right\}$ and $b_0 = \sum_{i=2}^{4} \frac{1}{2}(l_i^2 + \varepsilon_i^2) + \frac{m_1}{2r_1}\phi^2$. Furthermore, (12.27) implies that

$$V(t) \leq (V(t_0) - \frac{b_0}{a_0})e^{-a_0(t-t_0)} + \frac{b_0}{a_0} \leq V(t_0) + \frac{b_0}{a_0}, \forall t \geqslant t_0. \tag{12.28}$$

As a result, all $z_i (i = 1, 2, 3, 4)$ and $\tilde{\phi}$ belong to the compact set

$$\Omega = \left\{(z_i, \tilde{\phi}) | V \leq V(t_0) + \frac{b_0}{a_0}, \forall t \geqslant t_0\right\}.$$

Namely, all the signals in the closed-loop system are bounded. Especially, from (12.28) we have

$$\lim_{t \to \infty} z_1^2 \leq \frac{2b_0}{a_0}.$$

From the definitions of a_0 and b_0, it is clear that to get a small tracking error by taking r_i sufficiently large and l_i and ε_i small enough after giving the parameters k_i and m_i.

■

12.4 Simulation Results

In this section, we will compare the proposed method with the classical backstepping technique. To this end, the classical backstepping is first used to control design for the system (12.1), and the simulation is carried out by both of the proposed method and the classical one.

12.4.1 Classical Backstepping Design

The control of PMSM based on conventional backstepping technique is reviewed here.

Step 1: For the reference signal x_d, define the tracking error variable as $z_1 = x_1 - x_d$. From the first differential equation of (12.2), the error dynamic system is given by $\dot{z}_1 = x_2 - \dot{x}_d$.

Choose Lyapunov function candidate as $V_1 = \frac{1}{2}z_1^2$, then the time derivative of V_1 is computed by

$$\dot{V}_1 = z_1\dot{z}_1 = z_1(x_2 - \dot{x}_d). \tag{12.29}$$

Construct the virtual control law α_1 as

$$\alpha_1 = -k_1 z_1 + \dot{x}_d, \tag{12.30}$$

with $k_1 > 0$ being a design parameter and $z_2 = x_2 - \alpha_1$. By using (12.30), (12.29) can be rewritten of the following form.

$$\dot{V}_1 = -k_1 z_1^2 + z_1 z_2.$$

Step 2: Differentiating z_2 gives

$$\dot{z}_2 = \dot{x}_2 - \dot{\alpha}_1 = \sigma(x_3 - x_2) - \dot{\alpha}_1. \tag{12.31}$$

Now, choose the Lyapunov function candidate as $V_2 = V_1 + \frac{1}{2}z_2^2$. Obviously, the time derivative of V_2 is given by

$$\dot{V}_2 = \dot{V}_1 + z_2\dot{z}_2 = -k_1 z_1^2 + z_2(z_1 + \sigma(x_3 - x_2) - \dot{\alpha}_1). \tag{12.32}$$

The virtual control α_2 is constructed as

$$\alpha_2 = \frac{1}{\sigma}(-k_2 z_2 - z_1 + \sigma x_2 + \dot{\alpha}_1), \tag{12.33}$$

with $k_2 > 0$ being a design parameter and $\dot{\alpha}_1 = -k_1(\dot{x}_1 - \dot{x}_d) - \ddot{x}_d$. Adding and subtracting α_2 in the bracket in (12.31) shows that

$$\dot{V}_2 = \dot{V}_1 + z_2\dot{z}_2 = -k_1 z_1^2 - k_2 z_2^2 + z_2\sigma z_3, \tag{12.34}$$

with $z_3 = x_3 - \alpha_2$.

Step 3: Differentiating z_3 results in the following differential equation

$$\dot{z}_3 = \dot{x}_3 - \dot{\alpha}_2 = -x_3 - x_2x_4 + \gamma x_2 + u_q - \dot{\alpha}_2.$$

Choose the Lyapunov function candidate as $V_3 = V_2 + \frac{1}{2}z_3^2$. Furthermore, differentiating V_3 yields

$$\dot{V}_3 = \dot{V}_2 + z_3\dot{z}_3 = \dot{V}_2 + z_3(-x_3 - x_2x_4 + \gamma x_2 + u_q - \dot{\alpha}_2), \tag{12.35}$$

where

$$\begin{aligned}\dot{\alpha}_2 &= \sum_{i=1}^{2} \frac{\partial \alpha_2}{\partial x_i}\dot{x}_i + \sum_{i=0}^{2} \frac{\partial \alpha_2}{\partial x_d^{(i)}} x_d^{(i+1)} \\ &= \frac{\partial \alpha_2}{\partial x_1}x_2 + \frac{\partial \alpha_2}{\partial x_2}\sigma\,(x_3 - x_2) + \sum_{i=0}^{2} \frac{\partial \alpha_2}{\partial x_d^{(i)}} x_d^{(i+1)}.\end{aligned}$$

And the control law u_q is designed as

$$\begin{aligned}u_q &= -k_3z_3 + x_3 + x_2x_4 - \gamma x_2 + \dot{\alpha}_2 \\ &= -k_3z_3 + x_3 + x_2x_4 - \gamma x_2 + \frac{\partial \alpha_2}{\partial x_1}x_2 \\ &\quad + \frac{\partial \alpha_2}{\partial x_2}\sigma\,(x_3 - x_2) + \sum_{i=0}^{2} \frac{\partial \alpha_2}{\partial x_d^{(i)}} x_d^{(i+1)},\end{aligned} \tag{12.36}$$

with $k_3 > 0$. Furthermore, using the equality (12.36), it can be verified easily that

$$\dot{V}_3 \leq -\sum_{i=1}^{3} k_i z_i^2. \tag{12.37}$$

Step 4: At this step, we will construct the control law u_d. To this end, define $z_4 = x_4$ and choose the following Lyapunov function candidate as $V_4 = V_3 + \frac{1}{2}z_4^2$. Then the derivative of V_4 is given by

$$\dot{V}_4 = \dot{V}_3 + z_4\dot{z}_4 \leq -\sum_{i=1}^{3} k_i z_i^2 + z_4(-x_4 + x_2x_3 + u_d). \tag{12.38}$$

Now design u_d as

$$u_d = -k_4z_4 + x_4 - x_2x_3, \tag{12.39}$$

with $k_4 > 0$.

12.4.2 Simulation

Simulation is carried out for three cases under the initial condition of $x_1 = x_2 = x_3 = x_4 = 0.01$ in order to illustrate the effectiveness of the proposed results. In the first case, we tested the chaotic PMSM drive system with $u_d = u_q = 0$, which are shown in Figs. 12.2, 12.3 and 12.4. It is clearly seen that chaos occurs without control input signals. For another two cases, the proposed adaptive fuzzy approach is used to control the chaotic PMSM system for different σ, γ and reference signals. For the second case, $\sigma = 5.46, \gamma = 14.93$ and the given reference signal is $x_d = 0.5\sin(t) + 0.5\sin(0.5t)$; and for the third case $\sigma = 5.56, \gamma = 30$ and $x_d = \sin(2t) + \sin(t)$. The control parameters are chosen as follows:

$$k_1 = k_2 = k_3 = k_4 = 16, r_1 = 5, m_1 = 0.05, l_2 = l_3 = l_4 = 1.$$

Figures 12.2, 12.3 and 12.4. display the chaos when the control input is not implemented to control the PMSM drive system. In order to suppress chaos, the adaptive fuzzy controllers are used for another two cases. Figures 12.5, 12.6 and 12.7 show the simulation results for the second case, where Fig. 12.5 shows the reference signal x_d and the θ curves, Figs. 12.6 and 12.7 show the curves of i_d, i_q, u_d, u_q. Figs. 12.8, 12.9 and 12.10 are the simulation results for the third case. From the simulations, it is seen clearly that the proposed controller can suppress the chaos in the PMSM drive system and achieved a good tracking performance.

Remark 12.6 In practical application, the chaotic behavior in PMSM is undesirable since it can extremely destroy the stabilization of the motor even induce drive system collapse. So the control is implemented as soon as the motor is drove to avoid appearing the chaotic behavior. Figures 12.5, 12.6, 12.7, 12.8, 12.9 and 12.10 figures are provided to demonstrate the effectiveness of our method. ♦

Remark 12.7 In this research, fuzzy logic systems are employed to approximate nonlinearities which include the unknown σ and γ. In order to demonstrate the effectiveness of the fuzzy logic systems, different values of the unknown σ and γ are choose in simulation. Figures 12.5, 12.6, 12.7 and 12.8 demonstrate its effectiveness and robustness against the parameter uncertainties in the chaotic drive system. ♦

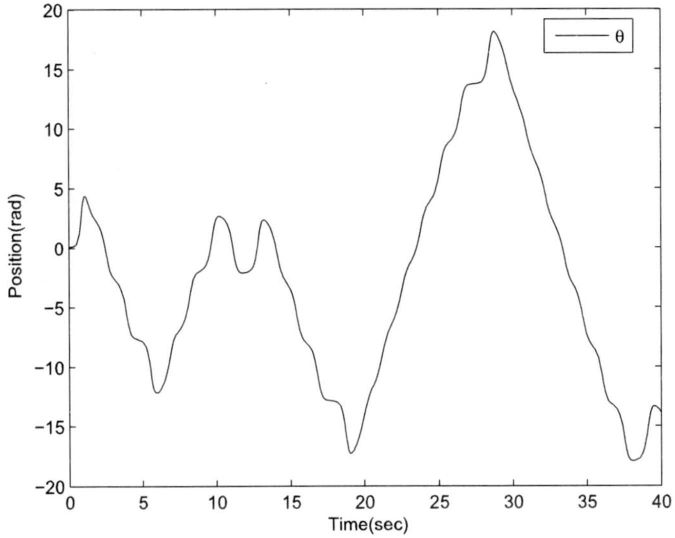

Fig. 12.2 Curve of the θ for chaotic PMSM system without u_d, u_q

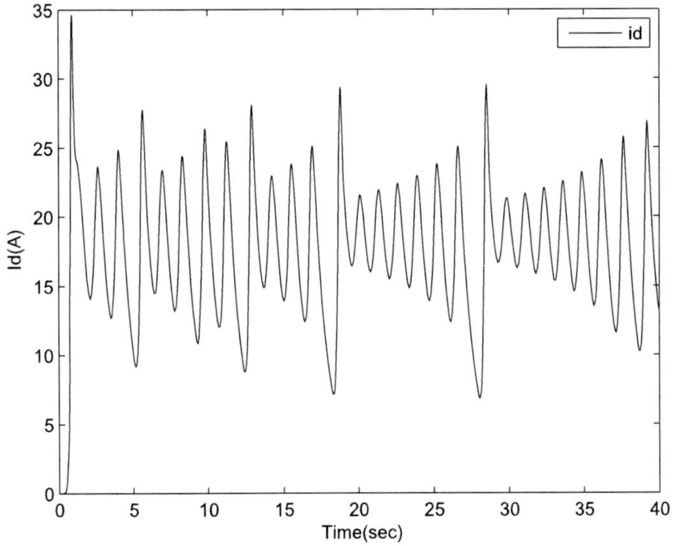

Fig. 12.3 Curve of the i_d for chaotic PMSM system without u_d, u_q

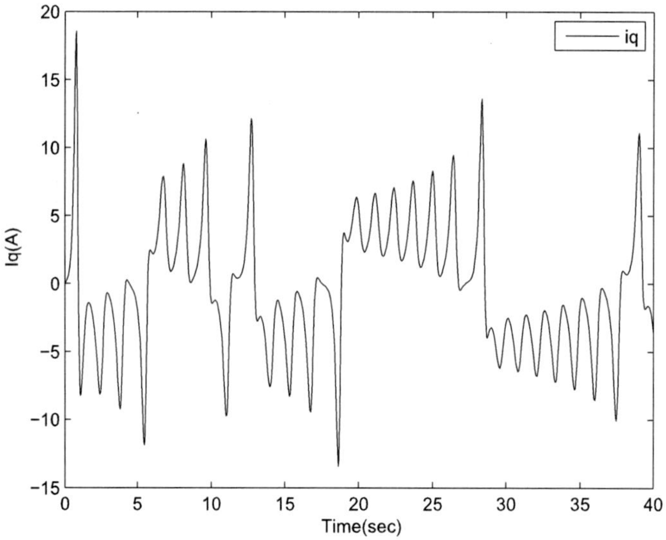

Fig. 12.4 Curve of the i_q for chaotic PMSM system without u_d, u_q

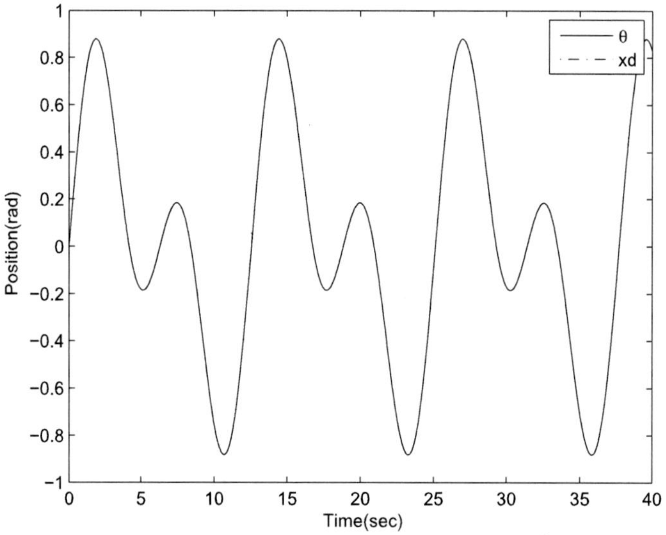

Fig. 12.5 Curves of the reference signal x_d and the θ for the second case

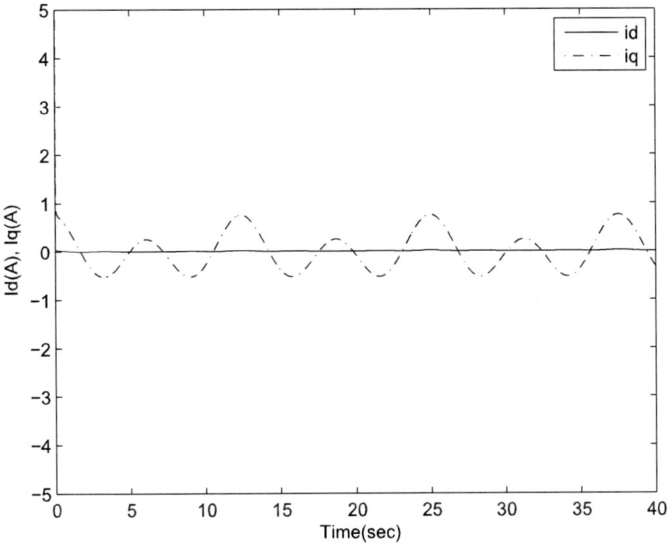

Fig. 12.6 Curves of the i_d and i_q for the second case

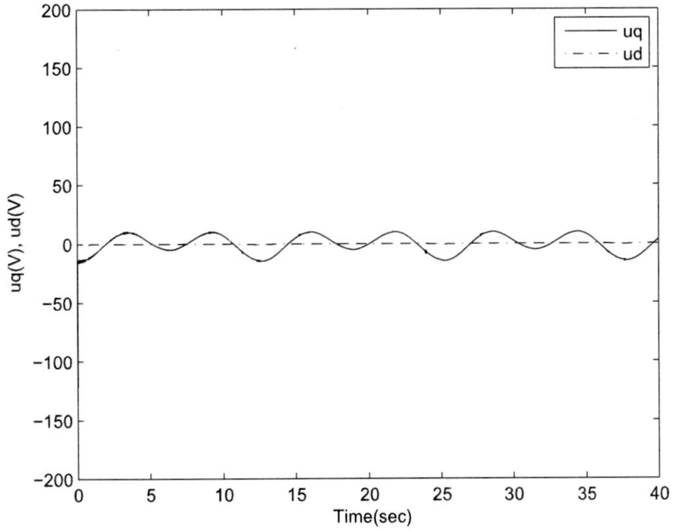

Fig. 12.7 Curves of the u_d and u_q for the second case

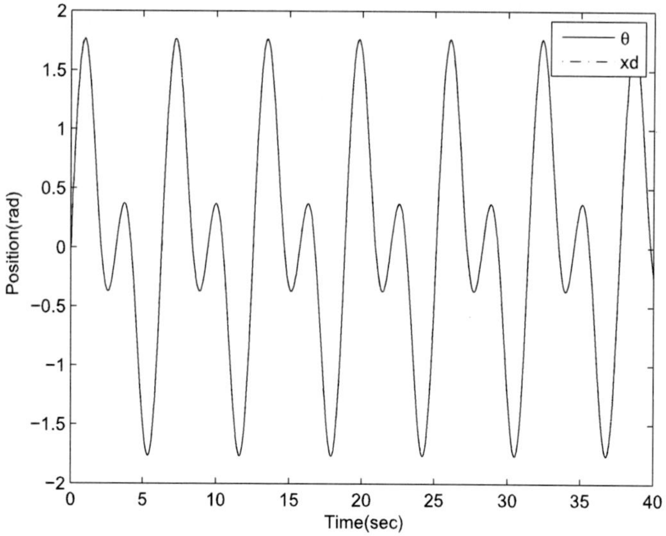

Fig. 12.8 Curves of the reference signal x_d and the θ for the third case

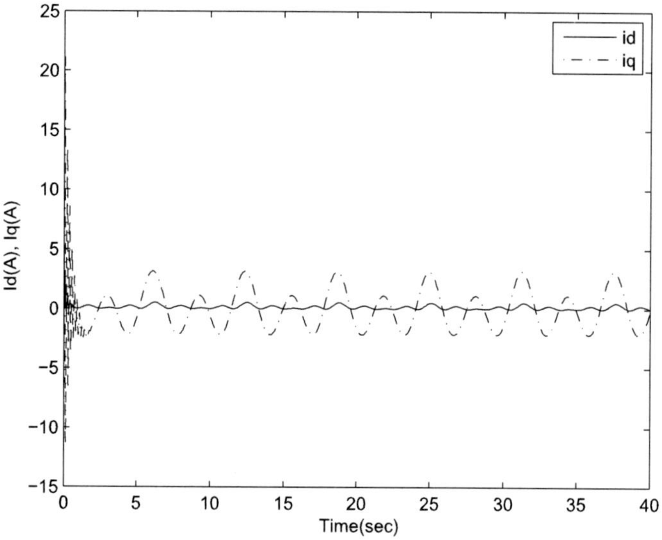

Fig. 12.9 Curves of the i_d and i_q for the third case

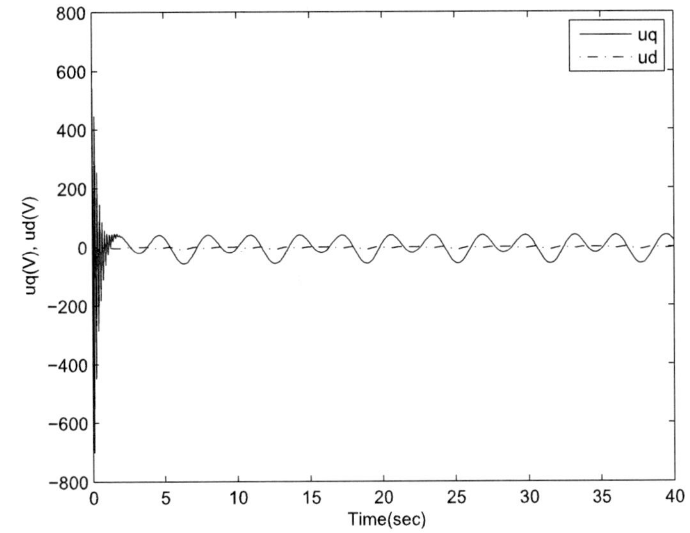

Fig. 12.10 Curves of the u_d and u_q for the third case

12.5 Conclusion

Aiming at the chaos problem in permanent magnet synchronous motor drive systems, an adaptive fuzzy control method based on backstepping technology is proposed. The controller overcomes the main problem of traditional backstep control, and ensures that the tracking error converges to a small neighborhood of the origin, and the closed-loop signal is bounded. Simulation results are provided to demonstrate the effectiveness and robustness against the parameter uncertainties in the chaotic drive system.

References

1. Boccaletti, S., Grebogi, C., Lai, Y.C., Mancini, H., Maza, D.: The control of chaos: theory and applications. Phys. Rep.-Rev. Sec. Phys. Lett. **329**, 103–197 (2000)
2. Liu, Y., Zheng, Y.: Adaptive robust fuzzy control for a class of uncertain chaotic systems. Nonlinear Dyn. **57**, 431–439 (2009)
3. Harb, A.M., Ahmad, W.A.: Control of chaotic oscillators using nonlinear recursive backstepping controllers. In: IASTED Conference on Applied Simulations and Modeling, pp. 451–453 (2002)
4. Kuroe, Y., Hayash, S.: Analysis of bifurcation in power electronic induction motor drive system. In: IEEE Power Electronics Specialists Conference, pp. 923–930 (1989)
5. Li, Z., Park, J.B., Joo, Y.H., Zhang, B., Chen, G.: Bifurcations and chaos in a permanent-magnet synchronous motor. IEEE Trans. Circuits Syst. I, Fundam. Theory Appl. **49**, 383–387 (2002)

6. Ren, H. and Liu, D.: Nonlinear feedback control of chaos in permanent magnet synchronous motor. IEEE Trans. Circuits Syst. II, Express Briefs **53**, 45–50 (2006)
7. Ren, H., Liu, D., Li, J.: Delay feedback control of chaos in permanent magnet synchronous motor. In: Proceedings of the China Society Electronic Engineering Conference, vol. 23, pp. 175–178 (2003)
8. Harb, A.M.: Nonlinear chaos control in a permanent magnet reluctance machine. Chaos Solitons Fractals **19**, 1217–1224 (2004)
9. Zribi, M., Oteafy, A., Smaoui, N.: Controlling chaos in the permanent magnet synchronous motor. Chaos Solitons Fractals **41**(3), 1266–1276 (2009)
10. Wei, D.Q., Luo, X.S., Wang, B.H., Fang, J.Q.: Robust adaptive dynamic surface control of chaos in permanent magnet synchronous motor. Phys. Lett. A **363**, 71–77 (2007)
11. Ge, X., Huang, J.: Chaos control of permanent magnet synchronous motor. In: Proceedings of the Eighth International Conference on Electrical Machines and Systems, vol. 1, pp. 484–488 (2005)
12. Wang, M., Liu, X., Shi, P.: Adaptive neural control of pure-feedback nonlinear time-delay systems via dynamic surface technique. IEEE Trans. Syst. Man Cybern. Part B Cybern. **41**(6), 1681–1692 (2011)
13. Tong, S.C., Zhang, W., Wang, T.: Robust stabilization conditions and observer-based controllers for fuzzy systems with input delay. Int. J. Innov. Comput. Inf. Control **6**(12), 5473–5484 (2010)
14. Wang, L.X., Ming, L.Y., Shi, P.: Fuzzy adaptive backstepping robust control for SISO nonlinear system with dynamic uncertainties. Inf. Sci. **179**(9), 1319–1332 (2009)
15. Wang, L.X., Mendel, J.M.: Fuzzy basis functions, universal approximation, and orthogonal least squares learning. IEEE Trans. Neural Netw. **3**(5), 807–814 (1992)
16. Stotsky, A., Hedrick, J., Yip, P.P.: The use of sliding modes to simplify the backstepping control method. In: Proceedings of the American Control Conference, vol. 3, pp. 1703–1708 (1997)

Part III
Summary

Chapter 13
Conclusion and Further Work

This chapter summarizes the results of the book and then proposes some related topics for future research.

13.1 Conclusion

The book mainly focusses on intelligent control problems for AC motors (including induction motor and permanent magnet synchronous motor). Specifically, detailed research problems have been listed as follows.

1. A new adaptive fuzzy control based on the backstepping technique has been designed for the position tracking of induction motor. Fuzzy logic systems are used to approximate the nonlinearities and an adaptive backstepping technique is employed to construct controllers. Lyapunov stability analysis shows that the proposed controller guarantees the tracking error converges to a small neighborhood of the origin.

2. A command filter adaptive neural-networks control scheme is proposed for the induction motor with input saturation. A smooth nonlinear function is introduced to deal with the nonlinearity caused by the input saturation. The command filtering technology is used to deal with the "explosion of complexity" problem caused by the derivative of virtual controllers in the conventional backstepping design.

3. A discrete-time neural networks controller based on the backstepping technique is designed for induction motor. The key problem in the discrete-time controller design process is the noncausal problem. The command filtered technique is introduced to eliminate the problem.

4. A new adaptive fuzzy control scheme is proposed for the induction motor with stochastic disturbances and input saturation. The quartic Lyapunov function is selected as the stochastic Lyapunov function and the adaptive backstepping method is used to design controllers. And then, the stability analysis is also given.

J. Yu et al., *Intelligent Backstepping Control for the Alternating-Current Drive Systems*,
Studies in Systems, Decision and Control 349,
https://doi.org/10.1007/978-3-030-67723-7_13

5. The influence of iron losses on the induction motor is considered. Firstly, A dynamic model of the induction motor considering iron losses is given. And then, the dynamic surface control is utilized to overcome the "explosion of complexity" problem of classical backstepping. Fuzzy logic systems are used to approximate unknown nonlinear functions and adaptive backstepping is employed to design controllers. Lyapunov stability analysis shows that the induction motor system is stable under the proposed control scheme.

6. An intelligent speed control is proposed for the PMSM with parameter uncertainties. Fuzzy logic systems are used to approximate nonlinearities and an adaptive backstepping technique is employed to construct controllers. Compared with the conventional backstepping method, the proposed fuzzy controllers' structure is very simple and easy to be implemented in practice.

7. Backstepping approach is utilized to solve the position tracking problem of the PMSM. By designing the fuzzy logic system and adaptive laws, a novel control scheme, which guarantees the tracking error converges to a small neighborhood of the origin and all the closed-loop signals are bounded, is obtained.

8. A neural-network-based adaptive dynamic surface control for the PMSM with parameter uncertainties and load torque disturbance is proposed. A novel adaptive dynamic surface control is constructed to avoid the explosion of complexity in the backstepping design and decreases the computation load of the control scheme. Stability analysis shows that the tracking error can converge to a small neighborhood of the origin.

9. A novel discrete-time adaptive position tracking controller is designed for the IPMSM. The noncausal problem in the discrete-time control is overcame by using the fuzzy logic systems. Stability analysis shows that all the state in the closed loop system are bounded under the proposed control scheme.

10. An adaptive fuzzy control method based on backstepping technology is developed to suppress chaos in the PMSM drive system. By employing the fuzzy logic systems, the proposed scheme can overcome the influence of uncertainty functions on the system performance.

11. A new intelligent control is proposed to overcome the influence of the chaos phenomena on the performance of the PMSM. A new dynamic model is used to characterize the chaos characteristics of the PMSM. By employing the fuzzy backstepping approach, a novel position tracking controller is proposed for the PMSM.

13.2 Further Work

Related topics for the future research work are listed below:

1. The adaptive fuzzy control method mainly relies on the structural information of the fuzzy basis functions. The fuzzy inference engine performs a mapping from fuzzy sets in Rnto fuzzy set in R based on the IF-THEN rules in the fuzzy rule base and the compositional rule of inference. How to choose the appropriate basis function vector and width coefficient so that the constructed adaptive fuzzy system can better

approximate the nonlinear function without increasing the burden of calculation is still a challenging question.

2. When the command filter is used to solve the computational complexity problem in the backstepping method, the filter will produce filtering error. The filter errors caused by the command filter would influence the performance of the closed loop system. However, the structure of the error compensation subsystem is still complex now. How to eliminate the filtering error in a simple way is the next work to be considered.

3. Motor drive systems inevitably need to consider the effects of full-state constraints. Meanwhile, actuator faults problems, such as short circuit or open circuit in power transistors, will cause catastrophic accidents. Hence, the full-state constraint and fault tolerant control problems will be considered in the next work.